DIGITAL TRANSFORMATION IN HEALTHCARE IN POST-COVID-19 TIMES

Next Generation Technology
Driven Personalized
Medicine and Smart Healthcare

DIGITAL TRANSFORMATION IN HEALTHCARE IN POST-COVID-19 TIMES

Edited by

MILTIADIS D. LYTRAS
*Deree College, The American College of Greece, Agia Paraskevi, Greece; Distinguished Scientists Program, Faculty of
Computing and Information Technology, King Abdulaziz University, Jeddah, Saudi Arabia;
Visiting Researcher, College of Engineering, Effat University, Jeddah, Saudi Arabia*

ABDULRAHMAN A. HOUSAWI
Saudi Commission for Health Specialties, Ministry of Health, Riyadh, Saudi Arabia

BASIM SALEH ALSAYWID
Education and Research Skills Directory, Saudi National Institute of Health, Riyadh, Saudi Arabia

ELSEVIER

ACADEMIC PRESS
An imprint of Elsevier

Academic Press is an imprint of Elsevier
125 London Wall, London EC2Y 5AS, United Kingdom
525 B Street, Suite 1650, San Diego, CA 92101, United States
50 Hampshire Street, 5th Floor, Cambridge, MA 02139, United States
The Boulevard, Langford Lane, Kidlington, Oxford OX5 1GB, United Kingdom

Notices

Knowledge and best practice in this field are constantly changing. As new research and experience broaden our
understanding, changes in research methods, professional practices, or medical treatment may become
necessary.

Practitioners and researchers must always rely on their own experience and knowledge in evaluating and using
any information, methods, compounds, or experiments described herein. In using such information or methods
they should be mindful of their own safety and the safety of others, including parties for whom they have a
professional responsibility.

To the fullest extent of the law, neither the Publisher nor the authors, contributors, or editors, assume any liability
for any injury and/or damage to persons or property as a matter of products liability, negligence or otherwise, or
from any use or operation of any methods, products, instructions, or ideas contained in the material herein.

ISBN: 978-0-323-98353-2

For information on all Academic Press publications
visit our website at https://www.elsevier.com/books-and-journals

Publisher: Stacy Masucci
Acquisitions Editor: Linda Versteeg-Buschman
Editorial Project Manager: Sara Pianavilla
Production Project Manager: Swapna Srinivasan
Cover Designer: Greg Harris

Typeset by STRAIVE, India

Working together
to grow libraries in
developing countries

www.elsevier.com • www.bookaid.org

Dedication

To all our colleagues and friends at the Saudi Commission for Health Specialties, the Saudi National Institute of Health, Effat University, and King Abdulaziz University for their continuous support, motivation, and inspiration to add value to our society through our hard work and to promote the implementation of Vision 2030 in Saudi Arabia with bold research and collaboration.

Dedication

Contents

A
Background of digital transformation in health

1. A historical outline of digital transformation

Cheng Gong and Vincent Ribiere

2. The potential of artificial intelligence in healthcare: Perceptions of healthcare practitioners and current adoption

Ahmed Hafez Mousa, Nishat Tasneem Maria, Fatimah Saleh Almuntashiri, Basim Saleh Alsaywid, and Miltiadis D. Lytras

B
Enabling technologies for digital transformation (an example-oriented approach)

3. Internet of things (IoT) and big data analytics (BDA) in healthcare

Prableen Kaur

4. Blockchain-based electronic health record system in the age of COVID-19

Yang Lu

C

Global experiences and approaches to digital transformation in health

Contributors

Haseena Al Katheeri Center for Student Success, Zayed University, Abu Dhabi, United Arab Emirates

Hasan Al-Attar Environment and Life Sciences Research Center, Kuwait Institute for Scientific Research, Safat, Kuwait

Ali Al-Hemoud Environment and Life Sciences Research Center, Kuwait Institute for Scientific Research, Safat, Kuwait

Yara Alkhalifah Fresno College of Public Health, California State University, Long Beach, CA, United States

Fatimah Saleh Almuntashiri Medicine and Surgery Program, Ibn Sina National Collage, Jeddah, Saudi Arabia

Basim Saleh Alsaywid Department of Urology, Pediatric Urology Section, King Faisal Specialist Hospital and Research Centre; Education and Research Skills Directory, Saudi National Institute of Health, Riyadh, Saudi Arabia

Nada A. AlTheyab Saudi Commission for Health Specialties, Riyadh, Saudi Arabia

Tayeb Brahimi Energy and Technology Research Lab, College of Engineering, Effat University, Jeddah, Saudi Arabia

Cheng Gong Institute for Knowledge and Innovation Southeast Asia, Bangkok University, Bangkok, Thailand

Meena Gupta Amity Institute of Physiotherapy, Amity University, Noida, Uttar Pradesh, India

Abdulrahman A. Housawi Saudi Commission for Health Specialties, Ministry of Health, Riyadh, Saudi Arabia

Fauzia Jabeen College of Business Administration, Abu Dhabi University, Abu Dhabi, United Arab Emirates

S. Sowmya Kamath Healthcare Analytics and Language Engineering Lab, Department of Information Technology, National Institute of Technology Karnataka, Surathkal, Mangaluru, India

K. Karthik Healthcare Analytics and Language Engineering Lab, Department of Information Technology, National Institute of Technology Karnataka, Surathkal, Mangaluru, India

Prableen Kaur Department of Mathematics, Chandigarh University, Chandigarh, Punjab, India

Prakash Kumar Amity Institute of Occupational Therapy, Amity University, Noida, Uttar Pradesh, India

Yang Lu School of Science, Technology and Health, York St John University, York, United Kingdom

Miltiadis D. Lytras College of Engineering, Effat University; Distinguished Scientists Program, Faculty of Computing and Information Technology, King Abdulaziz University, Jeddah, Saudi Arabia

Mariam Malek Environment and Life Sciences Research Center, Kuwait Institute for Scientific Research, Safat, Kuwait

Nishat Tasneem Maria Medicine and Surgery Program, Batterjee Medical College, Jeddah, Saudi Arabia

Veena Mayya Healthcare Analytics and Language Engineering Lab, Department of Information Technology, National Institute of Technology Karnataka, Surathkal, Mangaluru, India

Roberto Moro-Visconti Department of Economics and Business Management Sciences, Università Cattolica del Sacro Cuore, Milan, Italy

Ahmed Hafez Mousa Medicine and Surgery Program, Batterjee Medical College, Jeddah, Saudi Arabia

Divya Pandey Sai Institute of Paramedical and Allied Sciences, Dehradun, Uttar Pradesh, India

Jumanah Qedair College of Medicine, King Saud bin Abdulaziz University for Health Sciences, Jeddah, Saudi Arabia

Vincent Ribiere Institute for Knowledge and Innovation Southeast Asia, Bangkok University, Bangkok, Thailand

Ahmad Salman Kuwait Ministry of Health, Safat, Kuwait

Akila Sarirete Computer Science Department, College of Engineering, Effat University, Jeddah, Saudi Arabia

S.C.B. Samuel Anbu Selvan The American College, Affiliated to Madurai Kamaraj University, Madurai, Tamilnadu, India

Nazia Shehzad Faculty of Business, Liwa College of Technology, Abu Dhabi, United Arab Emirates

N. Vivek The American College, Affiliated to Madurai Kamaraj University, Madurai, Tamilnadu, India

Preface

The reactive nature of current healthcare systems globally forced their attention toward addressing acute and active chronic medical problems. As a result, healthcare has become practitioner and facility focused. Over time, care provision in such health systems has grown to be fragmented, inefficient, and expensive. Systems leaders must return the focus to people and populations to solve this challenge.

Health reform in the current era is characterized by shifting focus from volume-based to value-based care. As a result, health systems direct their resources and targets toward generating larger impacts on population health. Health systems must prioritize the screening, prevention, and early detection of chronic diseases to achieve this shift successfully. Furthermore, primary care becomes a central active component of new healthcare systems.

As a result of population expansion and vast geographic distributions, extending health services to remote areas beyond urbanized areas becomes necessary to transform health systems successfully.

Over the past three decades, parallel to changes in demands and priorities in healthcare systems, the world has been experiencing exponential technological advancement and growth. Subsequently, technology has been remarkably adopted and incorporated in different facets of healthcare systems, from operational and workflow enhancement to screening, diagnostics, and interventions.

The recent COVID-19 pandemic has accelerated technology adoption at rates that facilitated systems transformation on various fronts. Among the many examples of successful technology adoption within healthcare systems is the expansion in the utilization of remote health technologies that facilitated care provision to populations that would otherwise not be able to access healthcare services due to quarantine restrictions. It is important to note that adopting remote health technologies was facilitated by the coinciding application of health policies that support utilizing them.

Like with other industries, digital technologies and their accelerated advancement provide fertile soil for integrating technology within healthcare systems to address key issues (e.g., user focus, improved access, cost efficiency, etc.) in ways that will ultimately transform these systems. It is worth noting that many health systems and fields that realize this potential have demonstrated a remarkable transformation at faster rates than others. A unique feature of these pioneering systems is that their active technology adoption is based on a structured, visionary, and bold digital transformation strategy.

The utilization of digital technologies in health has always been limited by a myriad of potential risks, such as data privacy and cybersecurity. Therefore, for digital transformation to occur successfully, it is critical to address data privacy and cybersecurity as central aspects at all levels of technology adoption in healthcare systems. A formal and comprehensive digital transformation strategy best addresses such risks and potential challenges.

Several interrelated factors can explain the inconsistencies in digital transformation success between systems. Some of these factors are intrinsic to the health systems such as resource limitations, multiple competing priorities, and resistance to change while other factors are extrinsic, such as lacking regulatory and compliance policies.

Opportunity areas for digital transformation create an impact on the healthcare system

Healthcare systems are uniquely positioned to benefit from technologies to achieve transformative changes in several aspects. Patient experience is an important area where technology continues to improve, from scheduling to engagement and personalized experience. Remote health is an example of how technology can improve service provision and efficiency and the end user's perception of the services provided. This positive perception has improved the recipients' compliance and, ultimately, the quality of care.

Increasing costs constitute a major problem that taints healthcare delivery plans globally. The utilization of emerging technologies to address this complex issue has been shown to remedy numerous challenges previously deemed resistant to different interventions. One example is the use of intelligent systems to facilitate the diagnosis of specific medical problems with high degrees of accuracy, which is not possible with traditional approaches. The use of digital telepathology for populations that would not normally have access to pathology services highlights this benefit. The availability of digital telepathology allows for timely access to important diagnostic services at significantly lower costs, resulting in better care quality at a lower cost.

The potential of technology use and impact is demonstrable in many other domains of healthcare systems, including innovative clinical interventions, insurance claim management, inventory and asset management, health project management, and health system integration and interoperability.

Digital transformation from the healthcare practitioner perspective

The shift in healthcare system focus toward being more user-oriented calls for a paradigm shift in how the system operates. To successfully achieve this shift, a trait compatible with the shifting demands, there is a significant need for health system reform.

Value proposition of the book

This book discusses recent advances in patient care and offers critical comparative insights into their application across multiple domains in healthcare. By showcasing key problems, best practices, and emerging challenges, the book offers a state-of-the-art review of opportunities and prospects in delivering smart sustainable healthcare services. Topics discussed include healthcare challenges in the post-COVID-19 era; enabling technologies for digital transformation; value-driven approaches to the delivery of patient-centric top-quality health services; and analytics and enhanced decision making. In addition, the book updates knowledge on best practices for training toward digital transformation and sustainable health.

This is a valuable resource for healthcare professionals, medical doctors, researchers, graduate students, and members of the biomedical field who are interested in learning more about the use of emerging technologies in healthcare.

Organization of the book

The book is organized into 15 chapters under 3 sections:

Section A. Background of digital transformation in health.
Section B. Enabling technologies for digital transformation (an example-oriented approach).
Section C. Global experiences and approaches to digital transformation in health.

This three-tier approach helps the reader to understand the background knowledge and the foundations of digital transformation and enabling technologies. The strategic orientation of our volume is communicated in the third section where a systematic coverage of digital transformation strategies and best practices is delivered.

In most of the chapters, we deploy a bold active learning strategy. Readers are invited on exploratory journeys to the knowledge and domain supported by:

- Further readings
- Indicative assignments
- Short case studies
- Titles for essays
- Selected readings

From this point of view, this volume is a top-quality teaching reference for covering digital transformation in healthcare for diverse audiences:

1. Textbook for postgraduate courses and programs related to digital transformation in health.
2. Textbook for undergraduate courses in medical schools, information systems/ computer engineering departments, or policy-making departments.
3. Reference edition for health executive training programs covering holistically the theme of digital transformation in health.
4. Reference manual for professional development programs in healthcare management and innovation.

Coverage of the book

A historical outline of digital transformation

This chapter presents the history of digital transformation on a timeline basis illustrated by its key milestones from different perspectives, including etymological, the development of digital technologies, and the concepts' historical usage. The confusion and inconsistent use of digitization, digitalization, and digital transformation are discussed, and differentiating definitions are suggested following a rigorous scientific conceptualization process. To conclude, a brief introduction of digital transformation in healthcare is presented. The contribution of this chapter is a historical outline of digital transformation that depicts how digital-related terms have evolved throughout the past decades to clarify the vagueness and confusion in the field of digital transformation. Such an outline assuages the "fuzziness" issue associated with these concepts and inspires diverse communities and stakeholders in healthcare to use them more carefully, discriminatively, and consistently.

The potential of artificial intelligence in healthcare: Perceptions of healthcare practitioners and current adoption

The aim of this research is to assess and investigate the attitudes and perceptions of healthcare practitioners toward the application of artificial intelligence (AI) in the health field. An observational descriptive cross-sectional mapping study was organized in Saudi Arabia from June 6 to July 6, 2021. The study population and sampling included all healthcare practitioners from all specialties who agreed to be included in our study. AI has the potential to further improve patient care due to its ability to interpret more detailed and comprehensive data. As such, AI in medicine should be welcomed, especially after the power of various forms of AI has been proven in different aspects of healthcare, which is highly reflected in enhanced patient care and satisfaction. In this chapter, we provide initial findings, and we elaborate on some indicative aspects of the perceptions and the attitudes of healthcare practitioners for AI in Saudi Arabia.

Internet of things (IoT) and big data analytics (BDA) in healthcare

Enormous amounts of data are being generated daily and all this data are heterogeneous in nature. This heterogenous data can be beneficial for the healthcare sector when analyzed to get the value. Here, the data include medical prescriptions, laboratory data, electronic health records, and much more. These data are known as big data and they have a large handling capacity of computational analysis for modern computing systems. Also, for analysis purposes, the Internet of Things (IoT) acts as a major source for the data, revolutionizing the way we think about data. In this chapter, overviews of the IoT and big data analytics (BDA) in healthcare are presented. Emerging advances in IoT and big data analytics for healthcare technologies have helped in the coronavirus pandemic. Also, the challenges, applications, and the future in healthcare are

discussed. The paper also explores how the IoT and big data technologies can be combined for better healthcare solutions in this pandemic.

Blockchain-based electronic health record system in the age of COVID-19

The COVID-19 pandemic has exposed the limitations of current healthcare systems in responding to public health emergencies timely and efficiently. Regardless of the advantages of digitalized health records, a large portion of health systems are centralized and fall short in guaranteeing basic security and privacy requirements. As a decentralized and distributed technology, blockchain has been utilized for secure data access and information exchange in the healthcare domain. Moreover, COVID-19 health services such as antibody testing, vaccination proofing, contact tracing, health surveillance, and medical supply chains raise higher requirements in transparency, traceability, and immutability. Considering that the COVID-19 pandemic will have a long-term impact on all aspects of society, it is necessary to examine present developments for health information exchange. In this chapter, we discuss emerging platforms that adopt a blockchain architecture, with the focus on responses to the COVID-19 pandemic. Through a literature review, we show the critical role of blockchain technology in providing a more efficient, trustworthy, secure, and transparent health ecosystem. Finally, we identify future research directions along with recommendations.

Digital transformation in robotic rehabilitation and smart prosthetics

Ongoing improvement in smart prosthetic and rehabilitative gadgets has focused on further developing human-machine interactions.

In robotic rehabilitation, controlled exoskeletons and artificial limbs are intended to help individuals affected by spinal cord injury, amputation, and different neurodegenerative diseases. In present era, prosthetics and robotic rehabilitation are progressively ready to detect and think so they can interact with the human sensing environment and neural signaling with such ease prompting the neural movement reaction. Its myoneural interface empowers amputees to walk all the more normally and practically with no actual limitations. Mechanical restoration frequently incorporates and coordinates all the fundamental data from sensors such as EMG and touch sensors, encoding them and producing functional movements as required. Aside from this, it is planned by the adaptability of the clients. It examines the use of robotics in sensory as well as motor recovery and creates mechatronic help for developmental inabilities. In this section, we will focus on its function, method of utilization, and progress identified in robotic rehabilitation in the field of physical disabilities. This section will likewise include the role of advanced robotics for everyday life and how it can assist with the quality of living of the debilitated populace.

Ensemble deep neural models for automated abnormality detection and classification in precision care applications

Radiological imaging is one of the most relied upon modalities in the clinical diagnosis and treatment planning process. Conventional diagnosis involves the manual analysis of radiology images by experienced radiologists, which is often a time-consuming and labor-intensive process. The scarcity of experienced radiologists and necessity of largescale X-rays image analysis given the huge diagnosis workload at most hospitals stresses the need for automated clinical diagnosis systems capable of fast and accurate identification of abnormalities, disease characteristic identification, disease classification, and others. Such automated methods are thus a fundamental requirement in clinical workflow management applications. In this work, we present an approach for multitask clinical objectives such as disease classification and detection of abnormalities. The proposed model leverages the predictive power of deep neural models for enabling evidence-based diagnosis. During validation experiments, the model achieved an accuracy of 89.58% along with sensitivity and specificity of 85.83% and 90.83%, respectively, with an AUC (area under the ROC curve) of 95.84% for normal/no findings versus COVID-19 chest radiograph classification and an accuracy of 73.19% for upper extremity musculoskeletal images. The performance of the model for the classification and abnormality identification tasks, when benchmarked over multiple standard datasets, emphasizes its suitability and adaptability in real-world clinical settings, with significant improvements in radiology-based diagnosis workflow and patient care.

Smart healthcare digital transformation during the Covid-19 pandemic

The Covid-19 pandemic is revolutionizing our lives, forcing us to change our habits and reshaping our relationship with the healthcare system. Digitalization plays a crucial role, allowing for timely and aseptic connections that improve the efficiency of many healthcare processes. It shortens the supply chain—especially the last mile, from the hospital to the patient—and boosts the value chain that coalesces around value cocreating stakeholders, starting from the traditionally neglected patients.

The digital healthcare value chain starts from the Internet of Things and big data that are then stored in the cloud, interpreted with artificial intelligence patterns, and validated with blockchains. This chapter analyzes in an interdisciplinary way some practical applications of healthcare digital transformation that—thanks to networking digitalization—can become "smart," improving the quality of care and long-term socioeconomic sustainability.

Factors influencing the adoption of mobile health apps in the UAE

Increasing technology use has led to an increase in mobile health (mHealth) applications (apps). Users are adopting these applications for many reasons related to acceptance and gamification; however, there still needs to be a consensus on which factors most affect user adoption. Based on the Unified Theory of Acceptance and Use of Technology (UTAUT) model, this study examines the factors influencing the acceptance of mHealth technology. We collected data from 198 United Arab Emirate users who had previously used mHealth apps to improve their self-healthcare. The findings suggested that the intention of using this technology is positively influenced by: (1) the levels of performance expectancy and facilitating conditions concerning the use, (2) the level of gamification impact, and (3) the degree of personal innovativeness in the simple design of mHealth apps. These findings extend the theoretical concept of the UTAUT model in mobile healthcare technology. The value of the study lies in constructing an integrated digital transformation model that will assist healthcare

companies in comprehending the influence of available resources and lead them to invest significantly in the incumbent digital infrastructure. The findings shall help healthcare practitioners identify critical drivers and key challenges faced by stakeholders concerning mobile health app use in an emerging Arab country that has the vision to improve its healthcare through digital transformation.

Redesigning the global healthcare system through digital transformation: Insights from Saudi Arabia

According to the World Health Organization (WHO, 2010) "a health system consists of all the organizations, institutions, resources, and people whose primary purpose is to improve health." According to the same report, the building blocks of a modern health system consist of service delivery, health workforce, health information systems, access to essential medicines, financing, and leadership/governance.

A strategic framework for digital transformation in healthcare: Insights from the Saudi Commission for Health Specialties

This chapter elaborates on the key ideas of digital transformation as they are communicated in the recent literature. The authors synthesize complementary aspects for the study of the phenomenon and introduce a novel framework for a strategic approach to digital transformation in modern organizations. They also present the key determinants of digital transformation strategy with key emphasis on the Saudi Commission for Health Specialties. In their approach, three functional pillars of synergetic efficiency and performance for the digital transformation are highlighted: Pillar 1. Business Process Reengineering (Data, Knowledge, and Process Management); Pillar 2. Technology-Driven Value proposition for Digital Transformation; and Pillar 3. Quality Assurance (KPIs & Benchmarks). Additionally, the authors introduce five pillars of strategic aspects and impact for the digital transformation framework: Pillar 4. Policy Recommendations and Strategic Implications; Pillar 5. Digital Transformation Digital Innovation Foundations; Pillar 6. Value Creation and Cocreation; Pillar 7. Technology Enhancement; and Pillar 8. Platform-Mediated Ecosystems. Finally, the authors present the key challenges for the digital transformation of the Saudi Commission for Health Specialties in Saudi Arabia.

Digital transformation from a health professional practice and training perspective

This chapter discusses how Digital Transformation (DT) challenges health professional practice and training. A thorough argumentation on barriers, opportunities, and intersection areas of DT in the health domain sets the ground for understanding the wide context and application domain of DT in health organizations. The authors exploit further the key ideas and introduce significant use cases for DT projects in the Saudi Commission for Health Specialties (SCFHS). Special focus is paid to the core functional areas of the SCFHS and the ways that these can be digitally transformed. A pivotal project on stakeholder analysis and enhancement of value cocreation is also presented as an exemplar DT project in the SCFHS. The main ideas communicated in this

chapter can be used as a reference approach for potential DT projects in healthcare and health education organizations.

Research and Education Skills as a core part of Digital Transformation in Healthcare in Saudi Arabia

The adoption of digitalization technologies in healthcare systems is crucial for improving the delivery of medical services, facilitating patient access, and enhancing the overall patient experience. However, the experience and challenges of the Saudi healthcare system in adopting digital transformation have received little attention in the literature. Thus, this chapter aims to provide an overview of the digital transformation process in the Saudi healthcare system. The authors also shed light on the major challenges and suggested technology-based solutions. This chapter was written in light of the available global and local literature discussing the digitization of the health sector. Moreover, the Saudi governmental documents and official reports were utilized to include the latest updates and future plans for the local digital transformation strategy. In paving the way toward digital transformation, many challenges have been faced in areas such as medical records unification, technical infrastructure, and workforce capability. Solutions such as telemedicine applications, artificial intelligence, cybersecurity, and blockchain have been gradually applied to achieve the goal of transformation by 2030. Saudi Arabia's digital transformation began in 2018, with the implementation of the Vision 2030's rapid digital change. With the emergence of the COVID-19 pandemic, a leap in utilizing digital health solutions has been noticed locally. Nevertheless, more action plans are needed to address the challenges and implement suitable solutions accordingly.

A bibliometric analysis of GCC healthcare digital transformation

From diagnosing to preventing the spread of coronavirus, digital transformation and innovative technology have demonstrated their ability to play a key role in every aspect of the COVID-19 pandemic. Today, digital transformation goes beyond the application of artificial intelligence (AI) to increase productivity; it is currently reaching the large population in the Gulf Cooperation Council (GCC) and has a significant impact on both work and daily life. This study aims to evaluate the GCC's various contributions to scientific publications on digital transformation, focusing on the methods used to combat the COVID-19 pandemic and protect community well-being, including the most recent AI applications for COVID-19 safety measures, symptom detection, and remote healthcare. The research methodology used in this study is based on bibliometric analysis, a collection of strategies for analyzing vast amounts of bibliographic data by combining mathematical, statistical, and computer techniques. A set of publications is retrieved from three databases, Scopus, Web of Science (WoS), and lens.org. Then, VOS viewer is used to extract quantitative publication metrics and visualize coexisting networks of key terms extracted during the last 5years. This study focuses on the Scopus database while the WoS and the Lens databases are left for the user as an active learning process with some research directions in exploring bibliometric analysis in healthcare and digital transformation. From 2017 to 2021, 1520 healthcare, AI, and digital transformation documents were retrieved from the Scopus database using the journals' abstract, title, and keywords "TILE-ABS-KEY" sections. Results show that the total number of published documents in the GCC in healthcare and AI increased from 107 papers in 2017 to 720 papers in 2021. Furthermore, the number of citations jumped from 44 in 2017 to more than 4600 in 2021. The most active country was Saudi Arabia, followed by United Arab Emirates (UAE), Qatar, Oman, Kuwait, and Bahrain. Three of the top five most active institutions were from Saudi Arabia—King Saud University, King Abdulaziz University, and Imam Abdulrahman University—followed by the University of Sharjah from the UAE and Qatar University. Out of the 1520 documents retrieved, 20.6% were published in medicine, 18.6% in computer science, and 9.4% in engineering. Our findings indicate that AI and healthcare research are generally well established within each country, with more advancement in Saudi Arabia and UAE, but need more collaborative research between the GCC. This study provides a comprehensive overview of the bibliometric analysis of GCC healthcare digital transformation and AI, which may help researchers, policymakers, and practitioners better understand healthcare needs and development within the GCC.

Digital transformation of healthcare sector in India

Digital transformation is becoming a trend in the Indian healthcare sector. The healthcare sector in India has expanded in the last decade across both patient care as well as hospital services. The sector has invested in digital technologies such as social media platforms to associate with patients and develop relationships as well as in cloud platforms to perpetuate and keep track of health records and insurance transactions. The healthcare sector in India is starting to use analytics more rigorously for clinical and nonclinical data. Computerization solutions are still in the proof-of-concept stage and are mostly used in operations and other support functions but not instantaneously in medical functions. This chapter deals in detail with the foundation for the digital transformation, the need, the role of technology, the digital revolution, emerging trends, opportunities and challenges, future trends in healthcare digital transformation, and the like. This study would be a prototype to all the other upcoming studies in the aforementioned areas.

Investment in healthcare and medical technology in Kuwait

Although healthcare is the most promising field in the future for Kuwait, the country relies heavily on external resources for medical and healthcare technology. Kuwait's health sector lacks health knowledge and technical know-how. The investment in healthcare research is meager in Kuwait. One of Kuwait's future goals can be met by performing health research, arguably one of the most effective arenas for research. Advanced health research will aid in the establishment of a knowledge-based health industry in any country. Kuwait needs promising national research and development (R&D) efforts for promoting and managing sustainable development. Improvement in the health research area in Kuwait can significantly contribute toward improving overall public health, health promotion, disease prevention, and treatment. In Kuwait, applying current knowledge and skills through research has proved to be the most effective way of advancing

healthcare as well as getting extensively involved in translational medical research. In order to build an international medical hub, Kuwait needs to develop innovative medical treatment technology and a national health research complex. This study highlights the challenges that face the healthcare and medical technology in Kuwait.

Concluding remarks

This volume is developed with a great level of mutual respect and appreciation among all the contributors of the edition. It honors the great intellectual effort and the talent of numerous scholars from all over the world. We are really honored for their participation and great contribution.

This volume is not the end but the beginning of a great journey. We invite more colleagues to be on board. Currently, we are developing the continuation of this volume with a new title that will be available soon: **Next-Generation eHealth: Applied Data Science, Machine Learning, and Extreme Computational Intelligence**.

The editors of the book in collaboration with Elsevier will host an online section for learning and supportive material for this edition that will multiply the added value.

Miltiadis D. Lytras
Abdulrahman A. Housawi
Basim S. Alsaywid

Acknowledgment

With heartfelt thanks to the exceptional colleagues and contributors to this volume for their intellectual work and their commitment to a common vision for a better, happier, and healthier world for all.

Background of digital transformation in health

1

A historical outline of digital transformation

Cheng Gong and Vincent Ribiere

Institute for Knowledge and Innovation Southeast Asia, Bangkok University, Bangkok, Thailand

1. Introduction

The World Economic Forum (WEF, 2017) acknowledged digital transformation (DT) as one of the world's most pressing challenges regarding how various types of organizations can better meet evolving customer expectations, deliver their value propositions, and respond to a changing living and working environment. The growing penetration of digital technologies in the market inevitably drives the prevalence of DT, representing both disruptive difficulties and tremendous opportunities for the reinnovation of value offerings, business models, and organizational practices. While there is general agreement on its growing importance to an organization's success (Chanias and Hess, 2016; Sebastian et al., 2017), the inconsistent use of the term "digital transformation" in academia and business practice generates confusion.

On the academic front, the definitional inconsistency of DT and its related terms (e.g., digitization, digitalization) as well as the theoretical inconsistency of its implications at multiple levels of analysis hamper the betterment of research. The coexistence of numerous conflicting definitions has rendered these terms meaningless. This creates difficulties in developing a consistent stream of research that builds on what has been done before, thus making it more complicated to define and test relationships for DT theory building (Gong and Ribiere, 2021). The vagueness in the literature demonstrates the lack of a comprehensive, unified understanding of DT (Goerzig and Bauernhansl, 2018; Haffke et al., 2016; Matt et al., 2015; Morakanyane et al., 2017; Van Veldhoven and Vanthienen, 2019). This lack of a homogeneous interpretation of the concept is detrimental to research synergy, leading to wildly contradictory and incompatible research findings unfit to guide business practice.

On the practical front, DT appears to be one of the top priorities on business leaders' agendas (Sundblad, 2020). However, IBM research found that DT takes around 4 years to implement, but 85% fail (Gibson, 2018). A McKinsey study also revealed that the success rate for DT has been consistently low in recent years, as less than 30% succeed. Only 16% of organizations successfully improved performance, with an additional 7% of them saying that the improvements were not sustained in 2018. The success rates do not exceed 26% in digitally savvy industries (e.g., high tech, media, and telecom) and fall between 4% and 11% in more traditional industries (e.g., oil and gas, automotive, infrastructure, and pharmaceuticals) (De la Boutetière et al., 2018). Leaders and executives using the term DT inconsistently to describe various strategizing and organizing activities (Warner and Wäger, 2019) may risk blurring the distinct direction of organizational strategic moves (e.g., aiming for incremental vs. radical changes). Having an unclear DT vision challenges C-suite managers in claiming authority and clearly defining job responsibility for digital-related projects at the organizational level. Moreover, having diverse interpretations of DT makes it harder to benchmark one's performance against other organizations and industries on DT metrics and best practices at the industrial level.

Considering the "fuzziness" associated with DT and its related concepts in the extant literature, this chapter depicts the brief history of DT regarding its key milestones and players on a timeline basis, including a general discussion on the historical sequence of the most recognizable technologies in today's business landscape. How the diversity of these digital technologies broadens the impact of DT by creating network effects and combinatorial effects will be presented, followed by a detailed discussion of the historical use of digital-related concepts. The confusion and inconsistent use of digitization, digitalization, and DT in academia and practice are discussed, and differentiating definitions are provided based on a rigorous scientific conceptualization process. Diverse communities and stakeholders in healthcare are encouraged to rethink what they are and use them more carefully, discriminatively, and consistently for various strategic purposes in certain contexts, instead of simply using them interchangeably as buzzwords to attract attention.

2. The historical outline from different perspectives

2.1 From the etymological perspective

The words "digital" and "digitize" share a common Latin root: "digit," which emerged in ancient Latin as *digitus* (1st Century BC). This term originally meant "finger" or "toe" and evolved into the modern Latin word *digitalis* (since about 1500), meaning "fingers." According to Merriam-Webster (2022), the term "digit" was used as "any of the Arabic numerals 1 to 9 and usually the symbol 0" in the 14th century. It has been used in reference to a "numeral below 10" since the late 14th century. The term "digitize" was used as a verb to express the action "to finger, handle" from 1704, a sense now obsolete. Then, the modern use of the term "digital" as an adjective, meaning "of signals, information, or data: represented by series of discrete values (commonly the numbers 0 and 1), typically for electronic storage or processing" was in use from 1940 (OED, 2010).

George Stibitz first used this term in 1942 in the expression "digital computer" as a counterpart to the analog (Aspray, 2000). Thereafter, according to the Oxford English Dictionary, the term "digital" also means "of a computer or calculator: that operates on data in digital form; (of a storage medium) that stores digital data" (since 1945); and "of technologies, media, etc.: involving digital data; making use of digital computers or devices" (since 1948) (OED, 2010). These historical meanings of the word "digital" laid the foundation for the modern use of the verb "digitize," referring to "converting into a sequence of digits in computer programming, moving from analog number to electronic digits" (since 1953) (Online Etymology Dictionary, 2022).

Etymologically, the word "digitization" is rooted in the verb "digitize" while the word "digitalization" comes from the same Latin root "digital," which serves as one component of the concept "digital transformation." This etymological word commonality inevitably generates confusion between the meanings of these terms, which leads to an interchangeable use of these different terms in both academia and practice.

2.2 From the development of digital technologies perspective

In the past decades, the application of digital technologies has been widespread in many fields. The term "digital technologies" is therefore broadly used in different publications in reference to various technologies that are changing our daily lives and enabling DT. However, there is no consensus regarding what exactly these technologies are. They have basically been developing and emerging at an accelerating speed since the 1940s. Throughout these years, technological revolutions have changed the labor force in terms of creating new forms and patterns of work, leading to wider societal changes; meanwhile, they are also supporting the achievement of the United Nation's 17 sustainable development goals (UN, 2020).

After years of development, some of today's most advanced technological developments are expected to drive the next technological revolution, setting the scene for a wave of digitally enabled changes (Hoffmann and Ahlemann, 2019; Zhou, 2013). Although there are many fields where one unified standard for specific technology is not the outcome, as various standards can live side by side (Tadayoni et al., 2018), the tendency toward one or very few dominant standards can improve the compatibility of different standards so that an increased utility of the technology may be achieved. Taking mobile technologies as an example, the standardization processes of mobile networks illustrate the benefit of the so-called standard wars (i.e., battles for market dominance between incompatible technologies) (Shapiro and Varian, 1999a). Such processes happen along with each generation of mobile standards from 1G to 5G, and will perhaps impact future generations too. More specifically, on a timeline basis, mobile standards were either national or regional in the analog 1G phase in the 1980s, where handheld without wire networking used analog for phone calls solitary. A "standards war" between the newly developed digital mobile standards erupted with the 2G technologies in the 1990s, such as the European Community of the GSM (Groupe Spécial Mobile) standard and the prominent role of the European Telecommunications Standards Institute (ETSI), etc. The "standards war" continued with the development of 3G technologies in 2000, but relaxed with that of 4G technologies (Tadayoni et al., 2018); hence, some major competing standards were developed, such as LTE (by 3GPP), Mobile WiMAX (by IEEE), IMTs (by ITU), etc. The ongoing standardization processes of 5G technologies may make the outcome of the "standards war" more clear regarding cooperation among different organizations, nations, and regions (Sørensen et al., 2016). In related fields, smartphones and mobile apps have become a global norm since the outcome of a standards war. This is evident by the release of Apple's iPhone in 2007 and Google's Android-based smartphone in 2008 (Khansa et al., 2012). Other technologies, such as cloud computing, the Internet of Things (IoT), artificial intelligence (AI), big data analytics, three-dimensional (3D) printing, virtual reality (VR), augmented reality (AR), robotics, and blockchain, which are expected to become increasingly widespread in a range of industrial settings (Furman and Seamans, 2019), just tend to become new technological standards (Miric and Ozalp, 2020; Mukhopadhyay and Suryadevara, 2014; Rittinghouse and Ransome, 2016; Seaman, 2020).

There are 25 technologies mentioned in Gartner's 2021 Hype Cycle for Emerging Technologies (Panetta, 2021). These are well-recognized technologies that are commonly associated with digital technologies at the moment, and they are simply abbreviated as BCDIMR technologies[a] in this chapter. Beginning with Professor John McCarthy's suggestion that computing would be sold as a utility, then computer resource sharing through virtualization and the launch of the ARPANET (Advanced Research Projects Agency Network) since the 1960s, the foundational technologies for the cloud advanced over the next decades. Its reach to a certain level of maturity in the 1990s was epitomized by the launch of the worldwide web (WWW) in 1991, when more than a million machines were connected to the Internet (Varghese, 2019). Cloud computing first appeared in a Compaq internal document in 1996. Its inception was realized by the launch of the Amazon Web Service platform that consisted of only a few disparate tools and services in 2002 (AWS was officially relaunched in 2006 to provide an integrated suite of core online services). Since then, cloud computing has gone through two significant developments during the first generation regarding processing outside the cloud (2005–2011) and the second generation related to heterogeneous cloud resources (2012–2017) (Varghese, 2019). The IoT has come a long way, going from one or two machines in the 1980s (e.g., ARPANET-connected coke machine) to billions (Braun, 2019) with 620 IoT platforms in 2019 and $115 billion consumer spending on smart home systems worldwide in 2020 (Liu, 2020). The term "artificial intelligence" was coined by McCarthy in 1955 (Rajaraman, 2014). Then, various applications of AI have been refreshing the perception of what AI can do. For example, Unimate (the first industrial robot) went to work at General Motors (GM), replacing humans on the assembly line in 1961; Eliza (pioneering chatbot) came online in 1964; Deep Blue (a chess-playing computer) from IBM defeated world chess champion Garry Kasparov in 1997; Roomba (the first mass-produced autonomous robotic vacuum cleaner) came from iRobot in 2002; Siri (an intelligent virtual assistant with a voice interface) came from Apple; Watson (a question-answering computer) from IBM won first place on Jeopardy in 2011; Alexa (an intelligent virtual assistant with a voice interface that completes shopping tasks) came from Amazon in 2014; DuerOS (an intelligent voice assistant) came from Baidu in 2015; AlphaGo from Google beat world champion Ke Jie in the complex board game Go in 2017; and Waymo (the first fully self-driving taxi service legally allowed to operate) came about in 2018. Nowadays, AI and robotics are spreading like "technological wildfire" (De Miranda, 2019, p. 10). Over the last decade, cloud computing and its associated computational and storage capacities, the availability of big data, and breakthroughs in

---

[a] BCDIMR Technologies: B=Blockchain; C=Cloud computing; D=Big Data Analytics, 3D printing; I=Internet of Things (IoT), Artificial Intelligence (AI); M=Mobile technologies (5G/6G); R=Virtual Reality (VR), Augmented Reality (AR), Robotic process automation.

machine learning (ML) have effectually increased the power, availability, growth, and impact of AI (OECD, 2019). Although sociologist Charles Tilly (1984) first mentioned big data, Diebold (2012) argued that an unpublished 2001 research note by Douglas Laney (2001) at Gartner significantly enriched the concept. He is the person who originated the three Vs of the big data framework to describe big data in an up-to-date way (Egan and Haynes, 2019). This framework, which included volume, variety, and velocity (Chen et al., 2012; Russom, 2011), encapsulates the understanding of big data that refers to a large amount of data (both structured and unstructured) that arrives at high speed, often in real time (Egan and Haynes, 2019). With the development of social media, AI, and cloud technologies, businesses have implemented various advanced data analytics to take advantage of big data in recent years. Moreover, since the first patent was filed in the 1980s, many players have fought for the first-place position to become the brand name for 3D printing. Today, there are more than 170 3D printer system manufacturers worldwide, and purchasing a home 3D printer online is not hard (González, 2020). Still considered an emerging technology, VR/AR has come a long way over the last 50 years and has made big improvements in the past decades, mostly from the tech giant battle that ensued among Amazon, Apple, Alibaba, Google, Microsoft, Facebook, Sony, and Samsung (Poetker, 2019). It has been used in various industries over the years, from immersive marketing and entertainment to space missions. With the emergence of bitcoin and the development of Ethereum, numerous projects are exploring new features leveraging blockchain technology applications and addressing some of the security and scalability issues associated with early blockchain application (Iredale, 2018).

Overall, the advancement of digital technologies has expanded business's reach, allowing millions of people and business entities worldwide to connect and perform business functions (Terzi, 2011). The convergence of technologies will further take the expansion to the next level. Ray Kurzweil discovered in the 1990s that once a technology becomes digital or digitized (i.e., once it can be programmed in the 0s and 1s of computer code), it can hop on the back of Moore's Law and begin accelerating exponentially. "As exponential technologies converge, their potential for disruption increases in scale" (Diamandis and Kotler, 2020, p. 28). The diversity of digital technologies broadens the impact of DT by creating network effects (Fichman et al., 2014; Shapiro and Varian, 1999b) and combinatorial effects (WEF, 2017). Digital technologies are reflexive by nature (i.e., the use of one digital technology requires using other digital technologies), for instance, the use of a smartphone to utilize an app (Kallinikos et al., 2013). This reflexivity creates network effects arising from network externalities (Katz and Shapiro, 1985, 1986; Miyamoto, 2016; Pruett et al., 2003) and diffusion dynamics (Badiee and Ghazanfari, 2018), resulting in a lower cost and lowered entry barriers (Fichman et al., 2014). This way, the network efforts based on increased utility for the individual user with an increasing number of users and demand-side economies of scale can drive markets towards unified standards (Shapiro and Varian, 1999b; Tadayoni et al., 2018). This indicates that networks have an essential economic characteristic, that is, "the value of connecting to a network depends on the number of other people already connected to it" (Shapiro and Varian, 1999b, p. 175). Such a fundamental value proposition means in essence that it is better to be connected to a bigger network than a smaller one while other things are equal. This "bigger is better" aspect of networks that gives rise to positive feedback can be found under different names: network effects, network externalities, and demand-side economies of scale (Shapiro and Varian, 1999b). Meanwhile, ongoing technological progress is leading to better and cheaper sensors that capture more reliable data for use by AI systems and other technologies. The amount of data available for AI systems continues to grow as these sensors become smaller and less expensive to deploy (OECD, 2019). The progressive decrease (Matt, 2015) of the cost of some digital technologies (Saracco, 2017), increase in digital technology diversity (Routley, 2019), and the innovative use of digital technologies (Håkansson Lindqvist, 2019) are of great importance. On the one hand, they play a significant role in accelerating innovation as new applications; on the other hand, opportunities to combine different technologies in innovative ways are opened (WEF, 2017). This unleashes "combinatorial" effects, where what digital technologies working in tandem (i.e., the bundles of digital technologies) can achieve far exceeds what they can achieve when deployed separately, thus accelerating change and progress exponentially (WEF, 2017). Therefore, striving for network effects and combinatorial effects of digital technologies may offer tremendous potential beyond expectations.

2.3 From the concepts' historical usage perspective

The term "digital" has become popular along with the development of digital technologies mentioned in Section 2.2. In the past decades, different digital technologies have been researched and applied in diverse fields, giving rise to the popularity of using digital-related concepts (e.g., digitization, digitalization, and digital transformation). However, simply using these terms as buzzwords makes them meaningless and generates confusion. In contrast, understanding how they evolved and were used historically helps clarify this confusion, helping us to use them in a more meaningful and precise way.

2.3.1 Digitization

The Oxford English Dictionary (OED) traces the first modern use of the term "digitization" jointly with computers to the mid-1950s (OED, 2014). According to the OED, digitization refers to "the action or process of digitizing; the conversion of analog data (especially in later use images, video, and text) into digital form." Some scholars refer it to as the technical process of converting analog data into a digital format: an array of 0s and 1s stored in a way that makes them readable by computers. In fact, the use of this term is associated with the release of the first computer storage system based on magnetic disks (350 disk storage unit) incorporated in 305 RAMAC to store data in 1956. After that, the invention of the integrated data store (IDS) in 1964, one of the first database management systems, also strengthened the meaning of converting analog data into digital form for this concept.

With the release of several technological development milestones (see Fig. 1), the creation, storage, communication, and consumption of information and nondigital products are all being gradually digitized (Press, 2015). The first scanner, a drum scanner developed for use with a computer, was built in 1957. The first handwriting scanner—IBM 1287—was developed in the IBM Rochester lab in 1966. OCR-A and OCR-B typefaces made to facilitate optical character recognition (OCR) operations were introduced by American Type Founders and Swiss designer Adrian Frutiger in 1968. The first wide-area packet-switched network with distributed control—ARPANET—was born in 1969. Canadian postal operator Canada Post adopted

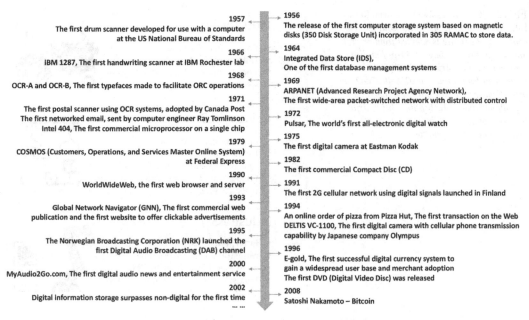

FIG. 1 The first milestones that reflect the history of digitization on a timeline basis.

the first postal scanner using OCR systems to read the name and address on the envelopes in 1971; in the same year, the first true microprocessor on a single chip—Intel 4004—was built and commercialized. The world's first all-electronic digital watch—Pulsar—appeared on the market in 1972. Despite the strategic failure of Kodak's choice to not explore the digital camera market, the first digital camera was indeed invented at Eastman Kodak in 1975. The first customers, operations, and services master online system (COSMOS) was used at Federal Express in 1979. The first commercial compact disc (CD) came out in 1982. The first web browser and server were created in 1990. The first 2G cellular network using digital signals launched in Finland in 1991 ended the analog 1G phase. The first commercial web publication and the first website to offer clickable advertisements—Global Network Navigator (GNN)—was introduced in 1993. The first digital transaction on the web was done for an online pizza order from Pizza Hut in 1994; in the same year, the world's first digital camera with cellular phone transmission capability—DELTIS VC-1100—was released. The Norwegian Broadcasting Corporation launched the first digital audio broadcasting (DAB) channel (NRK) in 1995. The first DVD was developed in 1995 and finally released in late 1996. The first successful digital currency system—E-gold—was designed to gain a widespread user base and merchant adoption in 1996. i2Go launched the first digital audio news and entertainment service called MyAudio2Go.com, which enabled users to download news, sports, and music in audio format in 2000. Digital information storage surpassed nondigital for the first time in 2002. A nine-page paper explaining a system of "electronic cash," authored by Satoshi Nakamoto (the presumed pseudonymous person or persons) was published in 2008, creating and deploying bitcoin's original reference implementation.

A. Background of digital transformation in health

FIG. 2 The successive Industrial Revolutions. *Adapted from online sources. The content of the figure is adapted from several online sources (e.g., https://medium. com/@winix/industry-4-0-the-digital-technology-transformation-b23ba02a7dd2; https://marine-offshore.bureauveritas.com/ digital/maritime-industry-40; http://drjiw. blogspot.com/2016/08/industry-40.html) and compiled by the authors.*

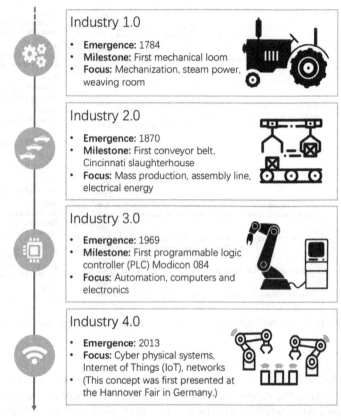

Industry 1.0
- **Emergence:** 1784
- **Milestone:** First mechanical loom
- **Focus:** Mechanization, steam power, weaving room

Industry 2.0
- **Emergence:** 1870
- **Milestone:** First conveyor belt, Cincinnati slaughterhouse
- **Focus:** Mass production, assembly line, electrical energy

Industry 3.0
- **Emergence:** 1969
- **Milestone:** First programmable logic controller (PLC) Modicon 084
- **Focus:** Automation, computers and electronics

Industry 4.0
- **Emergence:** 2013
- **Focus:** Cyber physical systems, Internet of Things (IoT), networks
- (This concept was first presented at the Hannover Fair in Germany.)

All these milestones in the development of digital technologies and their implications in different fields have compelled academics and practitioners to explore digital technologies' potential, extending from the technical process to their impact on different entities (i.e., organizations, businesses, industries, societies). Such achievements also led people to rethink how daily routines can be performed digitally, leading to the concept of a "paperless office," which was popular for some time to try to "digitize everything" (Tyler, 2021) or digitize as many documents as possible and reduce paper use. As a result, all these achievements contribute to the popularity of using the term "digitization" and gave rise to the use of other digital-related terms in terms of the impact beyond the technical process.

2.3.2 Digitalization

The first contemporary use of the term "digitalization" along with computerization appeared in Wachal's (1971) essay that discusses the social implications of the digitalization of society in computer-assisted humanities research (Brennen and Kreiss, 2016). In general, digitalization refers to "the use of digital technologies" (Srai and Lorentz, 2019, p. 79). It "loses its more technical aspects to digitization while maintaining the vague ideas of restructuring social life or business, and all the normative connotations they entail" (Seibt et al., 2019, p. 10). Seibt et al. (2019) argued that the discussion around the digitalization of industry is a debate that got labeled "Industry 4.0." Speaking of Industry 4.0, the name itself already illustrates the progressive and successive Industrial Revolutions since 1784 (see discussion in Box 1 in detail). This concept is the most prominent field of the industrial application of digitization, digitalization, and automation (Schumacher et al., 2016). However, this does not mean that these concepts refer to the same reality/phenomenon. Bloomberg (2018) noted that "automation is a major part of the digitalization story, whether it be shifting work roles or transforming business processes generally" (p. 4). It is "the full or partial replacement of a function previously carried out by the human operator" (Parasuraman et al., 2000, p. 287). Dijk van Jan (2006) noted that digitalization "allows a considerable increase in the production, dispersion, and consumption of information and the signals of communication" (p. 193), and "produces a culture of speed because creative production is assisted by the power of accelerated processing and distribution in computers and networks" (p. 209).

Historically, digitalization is often used as a synonym for DT when describing changes brought about by the adoption of digital technologies in society and organizations. The implementation of IT tools/software in organizations, such as material requirements planning (MRP), manufacturing resource planning (MRP II), enterprise resource planning (ERP), and business process reengineering (BPR) leads to the first generation of digitalization processes. During the 1970s and 1980s, with computer hardware and software development, MRP and MRP II emerged, driven by the need for stronger integration between the functional enterprise silos, the suppliers, and the customers. From the 1990s, ERP (i.e., the adoption of standard software packages) and BPR (i.e., business management initiatives striving for process efficiency supported by IT) started to emerge and spread. ERP is a "framework for organizing, defining, and standardizing the business processes necessary to effectively plan and control an organization so the organization can use its internal knowledge to seek external

---

BOX 1

Industry 4.0

To date, four recognizable Industrial Revolutions in history have enabled fundamental changes in how we live and work worldwide. Starting in the United Kingdom, the first Industrial Revolution with the invention and refinement of the steam engine and steam power enabled the mechanical production of goods. The second Industrial Revolution sparked developments in mass production. With the development of semiconductors and mainframe computers in the 1970s and the growth of the internet and the worldwide web in the 1990s, the third Industrial Revolution introduced the digital age internationally. Finally, using digital technologies presented in this chapter as well as the far-reaching mobile Internet combined with smaller, more powerful devices and sensors, the ongoing fourth Industrial Revolution is under way now. The following Fig. 2 summarizes key information about the successive Industrial Revolutions in history.

---

advantage" (Blackstone and Cox, 2005, p. 38). This dictionary definition resonates obviously with the expected outcomes of digitalization. The common aim/goal is to optimize organizations' existing business processes through efficient coordination among routines (Pagani and Pardo, 2017). Organizations may undertake a series of digitalization projects to automate processes and increase process efficiency (Bloomberg, 2018).

For practitioners, digitalization refers to "the use of digital technologies and data (digitized and natively digital) to create revenue, improve business, replace business processes (not simply digitizing them), and create an environment for digital business" (i-scoop, 2016), and "using digital technologies to automate processes for better outcomes and to optimize value" (NCMM, 2020). For scholars, digitalization refers to "the adoption of Internet-connected digital technologies and applications by companies" (Pagani and Pardo, 2017, p. 185), and "a means to fulfill customers' needs more effectively, adapt to changes in the sector, and increase their competitive advantage" (Rachinger et al., 2019, p. 1150).

In digitalization, digital technologies serve as enablers for organizations to improve and change their existing business processes (Verhoef et al., 2019), including communication (Ramaswamy and Ozcan, 2016; Van Doorn et al., 2010) and distribution (Leviäkangas, 2016). To achieve such goals, organizations may use ERP or other digital technologies to support the digitalization process. The changes ERP introduced are primarily limited to business processes within organizational boundaries in efficiency improvement, cost reduction, and business process optimization (Ash and Burn, 2003; Kauffman and Walden, 2001) because ERP mainly focuses on deploying internal management information systems (Boersma and Kingma, 2005). In a nutshell, ERP and BPR put effort into exploiting IT software packages and networking capabilities to improve organizational processes, focusing on production effectiveness and efficiency internally. Digitalization emphasizes the change process as a whole to achieve economic-driven outcomes through ERP or BPR and other digital technologies.

2.3.3 Digital transformation

There is no common consensus regarding the seminal scientific definition of DT in the literature. Historically, the ideas of digital products, services, and mediums can be traced back to the 1990s and 2000s (Auriga, 2016; Schallmo et al., 2017). Morton (1991) noted that organizations experience fundamental transformations for effective IT implementation. This idea gave birth to a research stream studying IT-enabled organizational transformation, which may be seen as one of the scholarly roots of DT research (Nadkarni and Prügl, 2020). It initiated DT's discussion with a strong IT focus as a catalyst of the information revolution (Gates et al., 1995) in the context of the information society's age and global competition. DT comprises some objectives and tasks that have been addressed in existing information management frameworks, which were subsumed under the prefix "digital," as well as methods and contents under the suffix "transformation" (Gong and Ribiere, 2021). Based on a research query on the SCOPUS and EBSCO databases, the first use of the term DT came in 2000.

At the early stage, a strong emphasis was put on the "digital" part—the use of digital technologies, providing a limited understanding of the "transformation" part of an entity. Thus, the concept of DT was often used, or probably misused, synonymously with that of digitization (the technical process) and digitalization (the installation process). With the accelerating development of digital technologies since the 1940s (see Fig. 3), industrial changes and

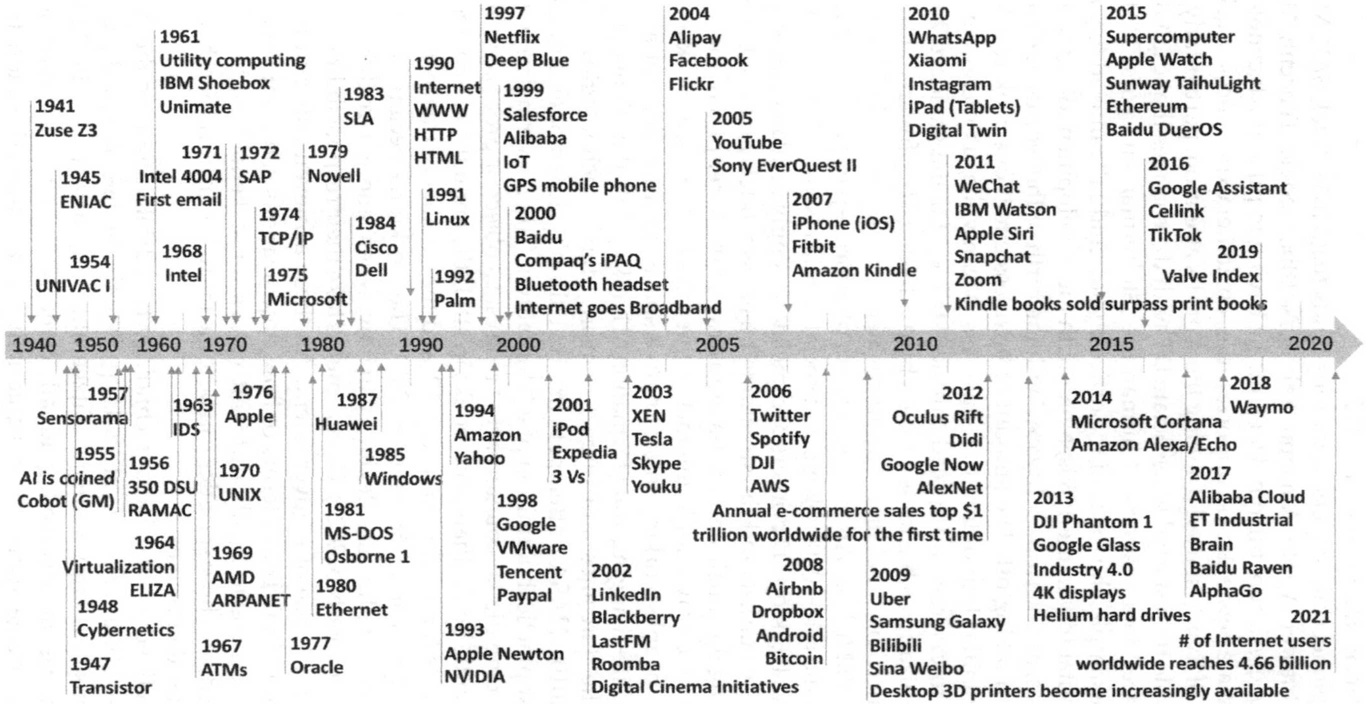

1941 Zuse Z3

1945 ENIAC

1954 UNIVAC I

1961 Utility computing
IBM Shoebox
Unimate

1947 Transistor

1948 Cybernetics

1964 Virtualization
ELIZA

1955 Cobot (GM)

1956 350 DSU
RAMAC

AI is coined

1957 Sensorama

1963 IDS

1971 Intel 4004
First email

1972 SAP

1974 TCP/IP

1975 Microsoft

1968 Intel

1969 AMD
ARPANET

1970 UNIX

1976 Apple

1967 ATMs

1979 Novell

1983 SLA

1984 Cisco
Dell

1987 Huawei

1985 Windows

1981 MS-DOS
Osborne 1

1980 Ethernet

1977 Oracle

1990 Internet
WWW
HTTP
HTML

1991 Linux

1992 Palm

1994 Amazon
Yahoo

1993 Apple Newton
NVIDIA

1997 Netflix
Deep Blue

1999 Salesforce
Alibaba
IoT

2000 Baidu
Compaq's iPAQ
Bluetooth headset

GPS mobile phone

Internet goes Broadband

1998 Google
VMware
Tencent
Paypal

2001 iPod
Expedia
3 Vs

2002 Linkedin
Blackberry
LastFM
Roomba

Digital Cinema Initiatives

2003 XEN
Tesla
Skype
Youku

2004 Alipay
Facebook
Flickr

2005 YouTube
Sony EverQuest II

2006 Twitter
Spotify
DJI
AWS

2007 iPhone (iOS)
Fitbit
Amazon Kindle

2008 Airbnb
Dropbox
Bitcoin

2009 Uber
Samsung Galaxy
Bilibili
Sina Weibo

Annual e-commerce sales top $1 trillion worldwide for the first time

Desktop 3D printers become increasingly available

2010 WhatsApp
Xiaomi
Instagram
iPad (Tablets)
Digital Twin

2011 WeChat
IBM Watson
Apple Siri
Snapchat
Zoom

2012 Oculus Rift
Didi
Google Now
AlexNet

2013 DJI Phantom 1
Google Glass
Industry 4.0
4K displays

2014 Microsoft Cortana
Amazon Alexa/Echo

Helium hard drives

Kindle books sold surpass print books

2015 Supercomputer
Apple Watch
Sunway TaihuLight
Ethereum
Baidu DuerOS

2016 Google Assistant
Cellink
TikTok

2017 Alibaba Cloud
ET Industrial Brain
Baidu Raven
AlphaGo

2018 Waymo

2019 Valve Index

2021 # of Internet users worldwide reaches 4.66 billion

1940 1950 1960 1970 1980 1990 2000 2005 2010 2015 2020

FIG. 3 Timeline of the key milestones and players in the history of DT.

societal developments throughout the previous decades could be seen, thus giving more importance to the transformational part of DT.

The efforts of pioneering players in various fields (e.g., IBM, Intel, SAP, Microsoft, Oracle, Cisco, Dell, Huawei, Apple, Amazon, Yahoo, Netflix, Google, Tencent, Alibaba, Baidu, Facebook, YouTube, Airbnb, Bitcoin, Uber, etc.), striving to push the technological revolution forward to create new business opportunities, customer experiences, working environments, and ways of living, started drawing the world's attention, particularly over the past two decades. People then started to associate DT with the changes that digital technologies caused or influenced in all aspects of human life (Stolterman and Fors, 2004). Taking the workplace as an example, the focus of the office layout gradually moved toward more productivity and profitability in the 1980s, along with the development of computers, portable computers, phones, and various software. With the birth of the worldwide web (WWW) in 1990, email started taking off. The amount of business email overtook regular mail in the 1990s, laying the initial foundation of e-business. Meanwhile, as mentioned earlier, ERP and BPR started to emerge and spread, further striking for the standardization and efficiency of an organization's business process by exploiting IT software. In the early 2000s, high-speed connectivity became hyperprevalent in society, and the pioneers of e-business were established. After several years of effort, e-payments started to be trusted and used around the mid-2000s. Skype and social media also made their way to the public around this period. Desktop 3D printers became increasingly available, the smartphone was brought to the mass market, and the streaming business models and digitized documentation started shaping the music and media industries in the latter part of the 2000s. All these changes redefined the way people interact with each other in their work and lives. It not only gives rise to people working in alternative spaces to the office but also puts people in the process of life-changing transformation. Thanks to all the pioneering players and early adopters, the outcomes and impact of using digital technologies continued to be enriching in the 2010s. NASA's attempt to improve the physical model simulation of spacecraft gave the first practical definition of a "digital twin" (i.e., a virtual representation that serves as the real-time digital counterpart of a physical object or process) in 2010. Amazon.com sold more Kindle books than print books in 2011; the annual e-commerce sales topped $1 trillion worldwide for the first time in 2012; and the number of Internet users worldwide reached 3 billion in 2014 and 4.66 billion in 2021. The workforce became much more mobile and began to demand agility and flexibility in how, when, and where they worked as technologies advanced (Condeco, 2022). Governments take their responsibilities to provide digital services, given the rise of e-government. All these achievements and transformations allowed people to rethink DT in terms of how organizations can completely change their business models and create totally different values for stakeholders, how industries can be radically shaped, and how societies can become more "smart," rather than only focusing on the technologies themselves.

DT has been defined as "the use of technology to radically improve the performance or reach of enterprise" (Westerman et al., 2011, p. 5) and "the use of new digital technologies to enable major business improvement" (Fitzgerald et al., 2014, p. 2). The "transformation" part of DT, which was undervalued, gradually came back to attention. As different research streams started to emerge, some scholars gradually realized that DT is more than just a technological shift (Henriette et al., 2015). Apart from technology, it requires "actors" (Nadkarni and Prügl, 2020) and the alignment of strategy and other factors, such as culture, mindset, talent development, and leadership (Goran et al., 2017).

In recent years, to get a comprehensive understanding of DT, some researchers have been concentrating on identifying DT's internal and external drivers, positive and negative consequences, etc. Diverse findings from different angles are emerging in the literature and seem to be enriched in the future too. For instance, external drivers may encompass: (1) innovation push and market pull generated by the adoption and development of digital technologies; (2) increasing volume of data; (3) accelerating customer behavior changes; and (4) laws/government policies adjustments. Internal drivers may include: (1) strategic imperatives (e.g., process and workplace improvement); (2) vertical and horizontal integration; (3) management support; and (4) cost reduction. Positive consequences may include: (1) decision-making improvement; (2) competitive advantage creation; and (3) value creation enhancement, (e.g., optimize customer experiences). Negative consequences may cover cybersecurity and privacy, etc. Beyond these research directions, debates regarding the true nature of DT and its essential dimensions are ongoing. As discussed by Gong and Ribiere (2021), the fundamental controversy may be founded in the fact that the range of DT definitions varies from a slight technology-enabled change (e.g., implementing a new ERP system) to a more radical and evolutionary process that takes place over time or the economic and societal effects of digitization and digitalization. In the extant literature, DT has been associated with business models and strategies (strategic moves) or viewed as a paradigm, process, etc. Consequently, the growing diversity of research fields associated with the concept of DT complexifies its clarification.

2.4 Synthesis

Historically, the terms digitization, digitalization, and DT are interconnected and describe different objects or phenomena. Digitalization, with a longer history of use in the literature than DT, inevitably encompasses the early discussion of digitization's social impact and the later discussion of digital transformation's result. A Google trend search by Seibt et al. (2019) indicates that the term digitization used to be more popular in English-speaking countries while the term digitalization has been more frequently searched for in continental Europe. This is still true as of today (see Fig. 4). In academia, the number of peer-reviewed publications (i.e., journal articles and conference papers) about DT considerably increased after the Industry 4.0 concept was presented at the Hannover Fair in Germany in 2013 (see Fig. 4). Around 95% of the publications appeared after 2013. Therefore, Industry 4.0 is inevitably associated with digitalization and DT in different geographical regions, research fields, and practices.

No distinction is widely represented in dictionaries, such as the Oxford dictionary, which offers the same definition for both terms. The Encyclopedia Britannica (Encyclopedia Britannica, 2022) and sociological dictionaries (Bruce and Yearley, 2006; Scott and Marshall, 2009; Swedberg and Agevall, 2016; Turner, 2006) do not define the terms digitization and digitalization. However, both terms are applied in business contexts and public debates by media (Seibt et al., 2019) with correlated meanings that have been causing a great deal of confusion. This indicates that there is no common consensus regarding the seminal scientific definition of these terms.

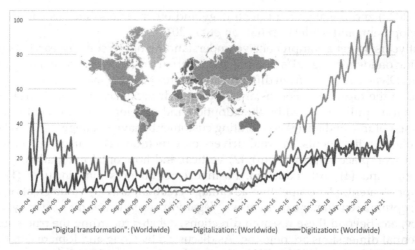

FIG. 4 The Google trend search of digital transformation, digitalization, and digitization. *Data source: https://www. google.com/trends.*

Digitization and digitalization are often applied to signify the same objects/phenomena. The same overlap exists between the use of digitization and DT. Some authors use different terms interchangeably, consciously or unconsciously; others may differentiate one concept while using the other two terms as equivalents implicitly or explicitly. Such confusion or lack of a common conceptual basis makes it impossible to ensure cumulative and sustainable knowledge creation (Sparrowe and Mayer, 2011). Consequently, this lack of clarity leads some authors to distinguish these three terms and their associated definitions in their articles to attach one specific term to one specific object/phenomenon (e.g., Mergel et al., 2019; Verhoef et al., 2019).

Apart from its truly intended meaning, digitalization has also been used to describe digitization in some cases and DT in other cases. Some authors, such as Verhoef et al. (2019), view the terms in sequential order (digitization→digitalization→digital transformation) with digitalization bridging and connecting the other two terms; other scholars disagree with this view. The situation is further complicated when linguistically translating digitalization and digital transformation as one word in some languages to explain the change and its end results of using digital technologies, not the technical process. Digitalization is used to depict a state of being digitalized and the process whereby the entities are affected by the action of "going digital."

Today's consensus seems to be that digital transformation is more than digitization (Haffke et al., 2016; Iansiti and Lakhani, 2014; Yoo et al., 2012). According to a review by Verhoef et al. (2019), most of the literature subscribes to the view that digitization and digitalization imply more incremental phases to attain the most pervasive phase of DT (Loebbecke and Picot, 2015; Parviainen et al., 2017). However, the inconsistent use of digitalization and DT still exists in a broad range of academic and practitioner literature. A disconcerting limitation of the existing literature is the failure to distinguish them properly.

2.5 Definition of digital-related terms

The previous section presented a historical outline of digital transformation from different perspectives: the etymological, the development of digital technologies, and the concepts' historical usage. It revealed that there is no common consensus regarding the seminal scientific definition of these terms. Confusion still exists in the extant literature regarding what they are and how to achieve them, as these terms are rooted in common language rather than in systematic scientific conceptualization. However, the use of these terms is popular in daily life as they are commonly encountered in different forms: advertisement, marketing, working, policy, etc.

Gong and Ribiere (2021) provided insights into core defining primitives of DT and linguistic clarity with a proposed unified definition to clarify such confusion. Based on their research, digital transformation can be defined as:

> A fundamental change process, enabled by the innovative use of digital technologies accompanied by the strategic leverage of key resources and capabilities, aiming to radically improve an entity [e.g., an organization, a business network, an industry, or society] and redefine its value proposition for its stakeholders (p. 12).

This unified DT definition aims to bring some conceptual rigor to DT at different levels in a broader context (e.g., organizational, industrial, and societal contexts) based on a vastly rigorous review and analysis of 134 well-received DT definitions in the literature. On the one hand, this unified definition improves DT's coherence by logically positioning its six primitives (i.e., nature, entity, means, expected outcome, impact, scope) to explain what DT is. On the other hand, it defines DT against related concepts by depicting what DT is and what DT is not to clarify the boundary of this concept.

Moreover, instead of focusing on DT definitions only, the following research conducted by Gong et al. (n.d.) further analyzed the logical alignment (i.e., internal coherence) and external differentiation of DT and its related concepts through a deep diachronic analysis of their defining attributes to make a clearer boundary for each concept to determine what reality is effectively attached to a particular concept. Consequently, distinctive definitions (in the organizational context) presented to differentiate them for the good of both academic and practitioner communities are offered as follows:

- *Digitization* is the technical process of converting analog into digital formats.
- *Digitalization* is the change process of installing digital technologies to reinforce the organization's existing value proposition.
- *Digital transformation* is a fundamental change process of an organization enabled by exploring the use of digital technologies to redefine its business models.

These definitions provide a clear distinction between these three concepts in the organizational context. Digitization refers to the technical process of converting analog into digital formats. The ultimate characteristic of being stripped of errors, repetition, and static allows digitized data and information to be easily stored, transferred, manipulated, and displayed, thus reducing paper clutter and improving efficiency. Digitalization and DT acknowledge the process enabled by digital technologies in organizations, yet they emphasize the different scope and end results of this process. Digitalization is implementing digital technologies to achieve economic-driven outcomes, that is, improve efficiency, productivity, and error elimination. It is a means to help the organization reinforce its existing value proposition efficiently and effectively, a change to the way of doing things. It does not involve a fundamental reappraisal of the organization's central assumptions or a paradigm shift of its organizational identity. In contrast, DT brings about digital innovation and transforms organizations' business operations to create new business models.

3. Digital transformation in different industries

In order to survive in a volatile, uncertain, complex and ambiguous world, well-established incumbents and new market entrants alike are increasingly seeking to leverage digital technologies to generate new revenue opportunities to differentiate in the market they compete in (Hoffmann and Ahlemann, 2019). There has been an expansion to identify and articulate unique aspects of digital technologies in various industries (Agrawal et al., 2015; Anderson and Agarwal, 2011), specific organizational domains (Xue et al., 2013), or product families (Greenstein et al., 2013) after witnessing several highly cited instances of digital disruption. Examples include the profound changes that Uber brings to taxis in the personal transportation sector (Berger et al., 2018), Airbnb to hospitality in the accommodation sector (Muller, 2020), and Amazon's Kindle to paper book in the publishing industry (Geissinger et al., 2020; Gilbert, 2015). Uber and Airbnb illustrate the effect of the sharing economy, which is a new industrial structure (Anderson and Huffman, 2017), in traditional sectors (Wallsten, 2015; Zervas et al., 2017). Amazon's Kindle exemplifies how the digitization of well-established products is fundamentally reshaping the structure that has underpinned book publishing for 200 years (Yoo et al., 2010). These changes in the industrial structure blur industry boundaries and create new opportunities and threats while bringing together organizations from previously unrelated industries (Yoo et al., 2010). Simultaneously, the broader options of diversified offerings (i.e., products and services) empowered by digital technologies are shifting power from organizations to customers. Rather than simply digitized products and services, increasing demand for high-quality experiences is evident. Customers tend to embrace offerings that deliver real value to them and apply these experiences to define their expectations across all other industries (WEF, 2017) (see Fig. 5). Advances in digital technologies are introducing new possibilities for hyperpersonalization in two forms: giving customers control to customize their product and experience, and providing more relevant interactions by analyzing customer data. Examples such as Nike's "customized shoes" and Brooks Brothers "create your own suit" build robust and agile supply chains by encouraging direct customer engagement at the point of sale. Other examples include Shopkick's personalized in-store shopping recommendations, and Ginger.io's mobile app, which use ML to monitor mental-health patients. Overall, the ubiquity and evolution of digital technologies have driven relentless changes in industrial structures, market dynamism, and customer preferences, and these changes have, in turn, placed new demands on organizations (Berman, 2012).

With the rise of on-demand healthcare, digital technologies are widely used in the healthcare sector. Examples include AI-enabled frontier technologies are helping to diagnose diseases and extend life expectancy; streamlined workflows using AI-powered systems; telemedicine; blockchain electronic health records; wearable medical devices; better treatment

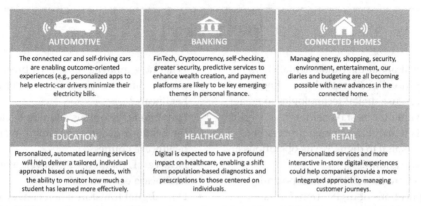

FIG. 5 The digital transformation in different industries. *Adapted from Digital Transformation Initiative, World Economic Forum Executive Summary, 2017.*

with VR tools; predictive healthcare using big data analytics, etc. In the healthcare sector, creating a rich health data foundation and integration of digital technologies are considered key components to tackle the challenges of improving patient outcomes while containing costs (Gopal et al., 2019). According to the Global Center for Digital Business Transformation)'s Digital Vortex 2021 report (Wade et al., 2021), healthcare is one of the four industries that showed the greatest digital acceleration from 2017 to 2021. A systematic literature review of the publication from 1973 to 2018 done by Marques and Ferreira (2020) also confirmed the growing trend of DT research in healthcare, especially after 1998. They identified seven categories: integrated management of IT in health, medical images, electronic health records, IT and portable devices in health, access to e-health, telemedicine, and privacy of medical data. Kraus et al. (2021) also conducted a systematic literature review to answer how multiple stakeholders implement digital technologies for management and business purposes. They found that prior research falls into five clusters: operational efficiency by healthcare providers; patient-centered approaches; organizational factors and managerial implications; workforce practices; and socioeconomic aspects. Different countries also settled the plan to take advantage of DT in their healthcare systems on the national level. For example, the Health Sector Transformation Program of Saudi Arabia (vision2030, 2021) aims to expand e-health services and digital solutions and improve healthcare quality in their Saudi Vision 2030 (Box 2).

BOX 2

Important questions to think about

- How to define "digital transformation" precisely and rigorously in healthcare?
- What does "digital transformation" mean in healthcare?

Is it the adoption of digital technologies in hospitals, physician practices, etc.? Is it the impact of digital technologies that affect the quality of care and efficiency?

Continued

A. Background of digital transformation in health

BOX 2   *(cont'd)*

Is it about the fundamental change process of a hospital regarding how it can use digital technologies to create value for its stakeholders?
Is it about the change process of a hospital to digitalize its operations and workflows?

- What are the effective and potential applications of digital transformation in healthcare?
- What are the fundamental changes that DT will bring to healthcare?

- What new innovative types of business models can we expect as a result of DT in healthcare?
- How can DT hospitals and other actors of the healthcare system become "smart healthcare entities"?
- How can DT support the development of healthcare ecosystems?
- Does digital transformation convey the same meaning of digitalization in healthcare?
- How can digital technologies be used to create network effects and combinatorial effects in healthcare?

---

4. Conclusion

Intending to get a more comprehensive understanding of how digital transformation has evolved, this chapter discussed the historical use of digital-related concepts on a timeline basis. First, we listed the confusion around DT on the academic and practical fronts, leading to the issue of "fuzziness" associated with DT and its related concepts in the extant literature. Second, we presented the historical outline of these concepts from different perspectives: the etymological, the development of digital technologies, and the concepts' historical usage. It revealed that there was no common consensus regarding the seminal scientific definition of these concepts. They are all rooted in common languages with the progress in the development of digital technologies rather than in systematic scientific conceptualization. We then discussed how the meanings of digitization, digitalization, and DT are interconnected and overlapped, leading to confusion and inconsistent use in academics and practice. Third, tackling the challenge of the multiplicity of their definitions, the authors' earlier research findings regarding the unified definition of DT and the distinctive definition of digitization, digitalization, and DT are provided. These definitions are based on rigorous scientific conceptualization processes and analyzing the existing defining attributes of extant definitions. They clarify what reality is effectively attached to a particular concept and the boundary for each concept. More importantly, we hope that they can serve as a reference to encourage readers to think about how DT can be defined scientifically and used carefully in healthcare. Finally, a general overview of DT in different industries and a brief discussion of DT in healthcare are presented to conclude this chapter.

In a nutshell, this chapter reviewed the literature on DT from various perspectives to provide insights into the essence of digital-related concepts. It contributes to building a common ground of how these concepts historically evolved so that researchers can build a consistent research stream based on what has been done before while practitioners can redefine their going digital strategies distinctively. After clarifying the confusion around digital-related concepts, the unified definition and distinctive definitions in the organizational context in Section 2.5 also offer new ways of thinking for diverse communities and stakeholders in healthcare (e.g., healthcare administrators, physicians, IT experts, policymakers, government officers, etc.) to reconsider how to strategically embark on their "going digital" journey. Different strategies and roadmaps need to be used in various contexts so that organizations and hospitals can fully take advantage of the resources efficiently and effectively. To our best understanding, there has been growing research interest in the digital healthcare transformation in recent years, especially after the COVID-19 pandemic. Much research is conducted to explore the application of different digital technologies to tackle the challenges of improving patient outcomes while containing costs. All in all, the historical outline of digital transformation depicts how digital-related terms have evolved throughout the past decades, clarifying the confusion around them for consistent research in the future.

References

Agrawal, A., Horton, J., Lacetera, N., Lyons, E., 2015. Digitization and the contract labor market: a research agenda. In: Economic Analysis of the Digital Economy. University of Chicago Press, pp. 219–250.

Anderson, C.L., Agarwal, R., 2011. The digitization of healthcare: boundary risks, emotion, and consumer willingness to disclose personal health information. Inf. Syst. Res. 22 (3), 469–490.

Anderson, M., Huffman, M., 2017. The sharing economy meets the sherman act: is uber a firm, a cartel, or something in between. Columbia Business Law Rev. 859.

Ash, C.G., Burn, J.M., 2003. Assessing the benefits from e-business transformation through effective enterprise management. Eur. J. Inf. Syst. 12 (4), 297–308.

Aspray, W., 2000. Computing Before Computers. Iowa State University Press, Ames, Iowa.

Auriga, 2016. Digital Transformation: History, Present, and Future Trends. Retrieved from: https://auriga.com/blog/digital-transformation-history-present-. and-future-trends/.

Badiee, A., Ghazanfari, M., 2018. A monopoly pricing model for diffusion maximization based on heterogeneous nodes and negative network externalities (case study: a novel product). Decision Sci. Lett. 7 (3), 287–300.

Berger, T., Chen, C., Frey, C.B., 2018. Drivers of disruption? Estimating the Uber effect. Eur. Econ. Rev. 110, 197–210.

Berman, S.J., 2012. Digital transformation: opportunities to create new business models. Strateg. Leadersh. 40 (2), 16–24.

Blackstone, J.H., Cox, J., 2005. APICS Dictionary (11E). APICS, Alexandria, VA.

Bloomberg, J., 2018. Digitization, Digitalization, and Digital Transformation: Confuse Them at Your Peril. Retrieved from: https://www.forbes.com/sites/jasonbloomberg/2018/04/29/digitization-digitalization-and-digital-transformation-confuse-them-at-your-peril/?sh=3ec17af32f2c.

Boersma, K., Kingma, S., 2005. From means to ends: the transformation of ERP in a manufacturing company. J. Strateg. Inf. Syst. 14 (2), 197–219.

Braun, A., 2019. History of IoT: A Timeline of Development. Retrieved from: https://www.iottechtrends.com/history-of-iot/.

Brennen, J.S., Kreiss, D., 2016. Digitalization. In: The International Encyclopedia of Communication Theory Philosophy, pp. 1–11.

Bruce, S., Yearley, S., 2006. The Sage Dictionary of Sociology. Sage.

Chanias, S., Hess, T., 2016. Understanding digital transformation strategy formation: insights from Europe's Automotive Industry. In: Paper presented at the Pacific Asia Conference on Information Systems (PACIS).

Chen, H., Chiang, R.H., Storey, V.C., 2012. Business intelligence and analytics: from big data to big impact. MIS Q., 1165–1188.

A. Background of digital transformation in health

Condeco, 2022. The History of the Workplace. Retrieved from: https://www.condecosoftware.com/modern-workplace/history-of-the-workplace/.

De la Boutetière, H., Montagner, A., Reich, A., 2018. Unlocking Success in Digital Transformations. Retrieved from: https://www.mckinsey.com/business-functions/organization/our-insights/unlocking-success-in-digital-transformations.

De Miranda, L., 2019. 30-Second AI & Robotics: 50 Key Notions, Fields, and Events in the Rise of Intelligent Machines, Each Explained in Half a Minute. Ivy Press.

Diamandis, P.H., Kotler, S., 2020. The Future Is Faster Than You Think: How Converging Technologies Are Transforming Business, Industries, and Our Lives. Simon & Schuster.

Diebold, F.X., 2012. On the origin (s) and development of the term 'Big Data'. In: Working Paper No. 12-037, Penn Institute for Economic Research, Philadelphia, PA.

Dijk van Jan, A., 2006. The Network Society. Social Aspects of New Media. SAGE Publications, London–Thousand Oaks–New Delhi.

Egan, D., Haynes, N.C., 2019. Manager perceptions of big data reliability in hotel revenue management decision making. Int. J. Quality Reliab. Manag. 36.

Encyclopedia Britannica, 2022. https://www.britannica.com/.

Fichman, R.G., Dos Santos, B.L., Zheng, Z., 2014. Digital innovation as a fundamental and powerful concept in the information systems curriculum. MIS Q. 38 (2), 329–A315.

Fitzgerald, M., Kruschwitz, N., Bonnet, D., Welch, M., 2014. Embracing digital technology: a new strategic imperative. MIT Sloan Manag. Rev. 55 (2), 1.

Furman, J., Seamans, R., 2019. AI and the economy. Innov. Policy Econ. 19 (1), 161–191.

Gates, B., Myhrvold, N., Rinearson, P., Domonkos, D., 1995. The Road Ahead. Viking Penguin, London.

Geissinger, A., Laurell, C., Sandström, C., 2020. Digital disruption beyond Uber and Airbnb—tracking the long tail of the sharing economy. Technol. Forecast. Social Change 155, 119323.

Gibson, D., 2018. Digital Transformation Takes Around Four Years and 85% of Them Fail, Says IBM. Retrieved from: https://www.thedrum.com/news/2018/12/04/digital-transformation-takes-around-four-years-and-85-them-fail-says-ibm-:~:text=Waite%20added%3A%20%E2%80%9CAny%20digital%20transformation,'t%20want%20to%20share.%E2%80%9D

Gilbert, R.J., 2015. E-books: a tale of digital disruption. J. Econ. Perspect. 29 (3), 165–184.

Goerzig, D., Bauernhansl, T., 2018. Enterprise architectures for the digital transformation in small and medium-sized enterprises. Proc. CIRP 67, 540–545.

Gong, C., Ribiere, V., 2021. Developing a unified definition of digital transformation. Technovation 102, 102217.

Gong, C., Parisot, X., Reis, D., n.d. Die evolution der digitalen transformation. In: Schallmo, D. (Ed.), DIGITALISIERUNG: Strategie, Transformation und Implementierung. Springer Nature, In press.

González, C.M., 2020. Infographic: The History of 3D Printing. Retrieved from: https://www.asme.org/topics-resources/content/infographic-the-history-of-3d-printing.

Gopal, G., Suter-Crazzolara, C., Toldo, L., Eberhardt, W., 2019. Digital transformation in healthcare—architectures of present and future information technologies. Clin. Chem. Lab. Med. 57 (3), 328–335.

Goran, J., LaBerge, L., Srinivasan, R., 2017. Culture for a digital age. McKinsey Quart. 3, 56–67.

Greenstein, M., Kennedy, C.B., Mikkelsen Jr., J.C., 2013. Automated System for Handling Microfluidic Devices. Google Patents.

Haffke, I., Kalgovas, B.J., Benlian, A., 2016. The role of the CIO and the CDO in an organization's digital transformation. In: Paper presented at the Thirty Seventh International Conference on Information Systems, Dublin.

Håkansson Lindqvist, M., 2019. School leaders' practices for innovative use of digital technologies in schools. Br. J. Educ. Technol. 50 (3), 1226–1240.

Henriette, E., Feki, M., Boughzala, I., 2015. The shape of digital transformation: a systematic literature review. In: MCIS 2015 Proceedings, pp. 431–443.

Hoffmann, D., Ahlemann, F., 2019. Harnessing Digital Enterprise Transformation Capabilities for Fundamental Strategic Changes: Research on Digital Innovation and Project Portfolio Management. Universität Duisburg-Essen.

Iansiti, M., Lakhani, K.R., 2014. Digital ubiquity: how connections, sensors, and data are revolutionizing business. Harv. Bus. Rev. 92 (11), 19.

Iredale, G., 2018. History of Blockchain Technology: A Detailed Guide. Retrieved from: https://101blockchains.com/history-of-blockchain-timeline/.

A. Background of digital transformation in health

i-scoop, 2016. Digitization, Digitalization and Digital Transformation: The Differences. Retrieved from: https://www.i-scoop.eu/digitization-digitalization-digital-transformation-disruption/.

Kallinikos, J., Aaltonen, A., Marton, A., 2013. The ambivalent ontology of digital artifacts. MIS Q., 357–370.

Katz, M.L., Shapiro, C., 1985. Network externalities, competition, and compatibility. Am. Econ. Rev. 75 (3), 424–440.

Katz, M.L., Shapiro, C., 1986. Technology adoption in the presence of network externalities. J. Polit. Econ. 94 (4), 822–841.

Kauffman, R.J., Walden, E.A., 2001. Economics and electronic commerce: survey and directions for research. Int. J. Electron. Commer. 5 (4), 5–116.

Khansa, L., Zobel, C.W., Goicochea, G., 2012. Creating a taxonomy for mobile commerce innovations using social network and cluster analyses. Int. J. Electron. Commer. 16 (4), 19–52.

Kraus, S., Schiavone, F., Pluzhnikova, A., Invernizzi, A.C., 2021. Digital transformation in healthcare: analyzing the current state-of-research. J. Bus. Res. 123, 557–567.

Laney, D., 2001. 3D data management: controlling data volume, velocity and variety. META Group Res. Note 6 (70), 1.

Leviäkangas, P., 2016. Digitalisation of Finland's transport sector. Technol. Soc. 47, 1–15.

Liu, S., 2020. Internet of Things (IoT)—Statistics and Facts. Retrieved from: https://www.statista.com/topics/2637/internet-of-things/.

Loebbecke, C., Picot, A., 2015. Reflections on societal and business model transformation arising from digitization and big data analytics: a research agenda. J. Strateg. Inf. Syst. 24 (3), 149–157.

Marques, I.C., Ferreira, J.J., 2020. Digital transformation in the area of health: systematic review of 45 years of evolution. Heal. Technol. 10 (3), 575–586.

Matt, R., 2015. Why Is Tech Getting Cheaper? Retrieved from: https://www.weforum.org/agenda/2015/10/why-is-tech-getting-cheaper/.

Matt, C., Hess, T., Benlian, A., 2015. Digital transformation strategies. Bus. Inf. Syst. Eng. 57 (5), 339–343.

Mergel, I., Edelmann, N., Haug, N., 2019. Defining digital transformation: results from expert interviews. Gov. Inf. Q. 36 (4), 101385.

Merriam-Webster (Ed.), 2022. Merriam-Webster.com Dictionary.

Miric, M., Ozalp, H., 2020. Standardized Tools and the Generalizability of Human Capital: The Impact of Standardized Technologies on Employee Mobility. Available at SSRN 3554224.

Miyamoto, M., 2016. An empirical examination of direct and indirect network externalities of the Japanese handheld computer industry: an empirical study of the early days. In: Wireless Communications, Networking and Applications. Springer, pp. 59–71.

Morakanyane, R., Grace, A.A., O'Reilly, P., 2017. Conceptualizing Digital Transformation in Business Organizations: A Systematic Review of Literature. Paper presented at the Bled eConference.

Morton, M.S., 1991. Corporation of the 1990s: Information Technology and Organizational Transformation. Oxford University Press, Inc, New York.

Mukhopadhyay, S.C., Suryadevara, N.K., 2014. Internet of things: challenges and opportunities. In: Internet of Things. Springer, pp. 1–17.

Muller, E., 2020. Delimiting disruption: why Uber is disruptive, but Airbnb is not. Int. J. Res. Mark. 37 (1), 43–55.

Nadkarni, S., Prügl, R., 2020. Digital transformation: a review, synthesis and opportunities for future research. Manag. Rev. Quart., 1–109.

NCMM, 2020. Digital Transformation Creates Middle Market Growth and Opportunity.

OECD, 2019. Artifical Intelligence in Society.

OED, 2010. Oxford English Dictionary, third ed. https://www.oed.com/viewdictionaryentry/Entry/52611.

OED (Ed.), 2014. The Oxford English Dictionary.

Online Etymology Dictionary, 2022. http://www.etymonline.com/.

Pagani, M., Pardo, C., 2017. The impact of digital technology on relationships in a business network. Ind. Mark. Manag. 67, 185–192.

Panetta, K., 2021. 3 Themes Surface in the 2021 Hype Cycle for Emerging Technologies. Retrieved from: https://www.gartner.com/smarterwithgartner/3-themes-surface-in-the-2021-hype-cycle-for-emerging-technologies.

Parasuraman, R., Sheridan, T.B., Wickens, C.D., 2000. A model for types and levels of human interaction with automation. IEEE Trans. Syst. Man Cybernet.-Part A: Syst. Humans 30 (3), 286–297.

Parviainen, P., Tihinen, M., Kääriäinen, J., Teppola, S., 2017. Tackling the digitalization challenge: how to benefit from digitalization in practice. Int. J. Inf. Syst. Proj. Manag. 5 (1), 63–77. https://doi.org/10.12821/ijispm050104.

A. Background of digital transformation in health

Poetker, B., 2019. The Very Real History of Virtual Reality (+A Look Ahead). Retrieved from: https://learn.g2.com/history-of-virtual-reality.

Press, G., 2015. A Very Short History of Digitization. Retrieved from: https://www.forbes.com/sites/gilpress/2015/12/27/a-very-short-history-of-digitization/?sh=4e0189e949ac.

Pruett, M., Lee, H., Lee, J.-R., O'Neal, D., 2003. How high-technology start-up firms may overcome direct and indirect network externalities. Int. J. IT Stand. Stand. Res. 1 (1), 33–45.

Rachinger, M., Rauter, R., Müller, C., Vorraber, W., Schirgi, E., 2019. Digitalization and its influence on business model innovation. J. Manuf. Technol. Manag. 30.

Rajaraman, V., 2014. JohnMcCarthy—father of artificial intelligence. Resonance 19 (3), 198–207.

Ramaswamy, V., Ozcan, K., 2016. Brand value co-creation in a digitalized world: an integrative framework and research implications. Int. J. Res. Mark. 33 (1), 93–106.

Rittinghouse, J.W., Ransome, J.F., 2016. Cloud Computing: Implementation, Management, and Security. CRC Press.

Routley, N., 2019. The Most Hyped Technology of Every Year From 2000-2018. Retrieved from: https://www.visualcapitalist.com/technology-hype-cycles-2000-2018/.

Russom, P., 2011. Big data analytics. TDWI Best Pract. Rep., Fourth Quarter 19 (4), 1–34.

Saracco, R., 2017. A Never Ending Decrease of Technology Cost. Retrieved from: https://cmte.ieee.org/futuredirections/2017/10/18/a-never-ending-decrease-of-technology-cost/.

Schallmo, D., Williams, C.A., Boardman, L., 2017. Digital transformation of business models—best practice, enablers, and roadmap. Int. J. Innov. Manag. 21 (08), 1740014.

Schumacher, A., Sihn, W., Erol, S., 2016. Automation, digitization and digitalization and their implications for manufacturing processes. In: Paper Presented at the Innovation and Sustainability Conference Bukarest.

Scott, J., Marshall, G., 2009. A Dictionary of Sociology. Oxford University Press, USA.

Seaman, J., 2020. China and the new geopolitics of technical standardization. French Inst. Int. Relat. 25 (03), 2020.

Sebastian, I.M., Ross, J.W., Beath, C., Mocker, M., Moloney, K.G., Fonstad, N.O., 2017. How big old companies navigate digital transformation. MIS Q. Exec. 16 (3), 197–213.

Seibt, D., Schaupp, S., Meyer, U., 2019. Toward an analytical understanding of domination and emancipation in digitalizing industries. In: Digitalization in Industry. Springer, pp. 1–25.

Shapiro, C., Varian, H.R., 1999a. The art of standards wars. Calif. Manag. Rev. 41 (2), 8–32.

Shapiro, C., Varian, H.R., 1999b. Information Rules: A Strategic Guide to the Network Economy. vol. 30 Harvard Business School Press, Boston, Massachusetts.

Sørensen, J., Tadayoni, R., Henten, A., 2016. 5G-Boundary object or battlefield? Commun. Strateg. 102, 63.

Sparrowe, R.T., Mayer, K.J., 2011. Publishing in AMJ—Part 4: Grounding Hypotheses. Academy of Management, Briarcliff Manor, NY.

Srai, J.S., Lorentz, H., 2019. Developing design principles for the digitalisation of purchasing and supply management. J. Purch. Supply Manag. 25 (1), 78–98.

Stolterman, E., Fors, A.C., 2004. Information technology and the good life. In: Information Systems Research. Springer, pp. 687–692.

Sundblad, W., 2020. CEO Insights: Will Digital Transformation Priorities Finally Stick? Retrieved from: https://www.forbes.com/sites/willemsundbladeurope/2020/07/08/ceo-insights-will-digital-transformation-priorities-stick/?sh=2ed5ff641f19.

Swedberg, R., Agevall, O., 2016. The Max Weber Dictionary: Key Words and Central Concepts. Stanford University Press.

Tadayoni, R., Henten, A., Sørensen, J., 2018. Mobile communications—on standards, classifications and generations. Telecommun. Policy 42 (3), 253–262.

Terzi, N., 2011. The impact of e-commerce on international trade and employment. Proc.-Social Behav. Sci. 24, 745–753.

Tilly, C., 1984. The old new social history and the new old social history. Review 7 (3), 363–406.

Turner, B.S., 2006. The Cambridge Dictionary of Sociology. Cambridge University Press.

Tyler, H., 2021. The Paperless Office: 35 Years of Dreaming Digital. Retrieved from: https://www.frevvo.com/blog/paperless-office/.

UN, 2020. Shaping Our Future Together. Retrieved from: https://www.un.org/en/un75/impact-digital-technologies.

A. Background of digital transformation in health

Van Doorn, J., Lemon, K.N., Mittal, V., Nass, S., Pick, D., Pirner, P., Verhoef, P.C., 2010. Customer engagement behavior: theoretical foundations and research directions. J. Serv. Res. 13 (3), 253–266.

Van Veldhoven, Z., Vanthienen, J., 2019. Designing a Comprehensive Understanding of Digital Transformation and its Impact. Paper presented at the Bled eConference.

Varghese, B., 2019. History of the Cloud. Retrieved from: https://www.bcs.org/content-hub/history-of-the-cloud/.

Verhoef, P.C., Broekhuizen, T., Bart, Y., Bhattacharya, A., Dong, J.Q., Fabian, N., Haenlein, M., 2019. Digital transformation: a multidisciplinary reflection and research agenda. J. Bus. Res. 112, 889–901.

vision2030, 2021. Health Sector Transformation Program. Kingdom of Saudi Arabia Retrieved from https://www.vision2030.gov.sa/v2030/vrps/hstp/.

Wachal, R., 1971. Humanities and computers: a personal view. North American Rev. 256 (1), 30–33.

Wade, M., Shan, J., Bjerkan, H., Yokoi, T., 2021. Digital Vortex 2021: Digital Disruption in a COVID World. Retrieved from: https://www.imd.org/contentassets/8c5b42807da941ee95c7be87d54e5db9/20210427-digitalvortex21-report-web-final.pdf.

Wallsten, S., 2015. The competitive effects of the sharing economy: how is Uber changing taxis. Technol. Policy Inst. 22, 1–21.

Warner, K.S.R., Wäger, M., 2019. Building dynamic capabilities for digital transformation: an ongoing process of strategic renewal. Long Range Plan. 52 (3), 326–349.

WEF, 2017. Digital Transformation Initiative: In Collaboration With Accenture. Retrieved from: http://reports.weforum.org/digital-transformation/wp-content/blogs.dir/94/mp/files/pages/files/dti-executive-summary-website-version.pdf.

Westerman, G., Calméjane, C., Bonnet, D., Ferraris, P., McAfee, A., 2011. Digital transformation: a roadmap for billion-dollar organizations. In: MIT Center for Digital Business Capgemini Consulting. vol. 1, pp. 1–68.

Xue, L., Zhang, C., Ling, H., Zhao, X., 2013. Risk mitigation in supply chain digitization: system modularity and information technology governance. J. Manag. Inf. Syst. 30 (1), 325–352.

Yoo, Y., Boland Jr., R.J., Lyytinen, K., Majchrzak, A., 2012. Organizing for innovation in the digitized world. Organ. Sci. 23 (5), 1398–1408.

Yoo, Y., Henfridsson, O., Lyytinen, K., 2010. Research commentary—the new organizing logic of digital innovation: an agenda for information systems research. Inf. Syst. Res. 21 (4), 724–735.

Zervas, G., Proserpio, D., Byers, J.W., 2017. The rise of the sharing economy: estimating the impact of Airbnb on the hotel industry. J. Mark. Res. 54 (5), 687–705.

Zhou, J., 2013. Digitalization and intelligentization of manufacturing industry. Adv. Manuf. 1 (1), 1–7.

A. Background of digital transformation in health

The potential of artificial intelligence in healthcare: Perceptions of healthcare practitioners and current adoption

Ahmed Hafez Mousa[a], Nishat Tasneem Maria[a], Fatimah Saleh Almuntashiri[b], Basim Saleh Alsaywid[c,d], and Miltiadis D. Lytras[e,f]

[a]Medicine and Surgery Program, Batterjee Medical College, Jeddah, Saudi Arabia [b]Medicine and Surgery Program, Ibn Sina National Collage, Jeddah, Saudi Arabia [c]Department of Urology, Pediatric Urology Section, King Faisal Specialist Hospital and Research Centre, Riyadh, Saudi Arabia [d]Education and Research Skills Directory, Saudi National Institute of Health, Riyadh, Saudi Arabia [e]College of Engineering, Effat University, Jeddah, Saudi Arabia [f]Distinguished Scientists Program, Faculty of Computing and Information Technology, King Abdulaziz University, Jeddah, Saudi Arabia

1. Introduction

According to Ellahham (2020), artificial intelligence (AI) has been described as "a branch of computer science that aims to create systems or methods that analyze information and allow the handling of complexity in a wide range of applications."

The significance of AI in the present era cannot be overstated. Simplified, AI is a program that attempts to mimic human cognitive function. Some examples of AI include Microsoft's Cortana, Amazon's Alexa, Apple's Siri, Google's assistant, product suggestions, self-driving cars, and facial recognition (Bohr and Memarzadeh, 2020).

Fear of change is not new, and neither is the concept of AI. However, public understanding of AI is still new. Eventually the discrepancy between the emergence and growth of AI versus adequate public knowledge will lead to concerns and dismay. Part of that fear is reasonable whereas the rest of it is due to a lack of adequate knowledge. This emergence has led to concerns as to whether healthcare workers would embrace this change.

The visceral fear of automation and being replaced by a machine is common. There is no doubt that AI has surpassed humans in many domains, especially in visual detection and identifying discrepancies (Ahuja, 2019).

AI has also begun to be incorporated into the medical field to improve care provided to patients by accelerating various processes and achieving greater accuracy, opening the path to providing enhanced healthcare overall. Radiological images, pathological slides, and patient electronic health records (EHRs) are being evaluated by machine learning, aiding in the diagnosis and treatment of patients and augmenting physician capabilities.

In April 2016, Saudi Arabia's Vision 2030 identified the long-term goals and expectations of the country. The Saudi Data and Artificial Intelligence Authority (SDAIA) was established in August 2019 to facilitate the transition and help achieve Vision 2030's goals. SDAIA's core mandate is to support and drive the data and AI agenda within the country, and its vision is to position Saudi Arabia as a global leader in the elite league of data-driven economies (Maliha et al., 2021). Our study aims to identify the current impact and assess the perspective of healthcare providers of AI in the near future (YESSER E-Government Program, n.d.).

References	Article title	Key contribution	Impact on our research model
1. Klaus Schwab	The Fourth Industrial Revolution	Purpose of the study was to investigate how revolution is emerging and creating a new core of society	Impact of revolution and widespread neural networking
2. Melissa D. McCarden	Conditionally positive: a qualitative study of public perceptions about using health data for artificial intelligence research	Given widespread interest in applying artificial intelligence AI to health data to improve patient care and health system efficiency, there is a need to understand the perspectives of the general public regarding the use of health data in AI research	Multidimensional perspective of patient, physician, and stakeholders regarding AI in application
3. Angeliki Kerasidu	Artificial intelligence and ongoing need for empathy, compassion, and trust in healthcare	To reevaluate whether and how values could be incorporated and practiced within a healthcare system where artificial intelligence is increasingly used. Most importantly, society needs to reexamine what kind of healthcare it ought to promote	Discussing the irreplaceable aspects of healthcare as well as how to ensure the healthy use of AI

A. Background of digital transformation in health

References	Article title	Key contribution	Impact on our research model
4. Soon Hyo Kwon		To investigate the awareness of AI among Korean doctors and to assess physician attitudes toward the medical application of AI	In case of radiology and pathology, some believe that AI will replace doctors based on diagnostic superiority
5. Varun Gulshan	Development and validation of a deep learning algorithm for the detection of diabetic retinopathy in retinal fundus photographs	To apply deep learning to create an algorithm for automated detection of diabetic retinopathy and diabetic macular edema in retinal fundus photographs	Artificial intelligence analytics can be used in chronic disease management characterized by multiorgan involvement and acute variable events. For instance, retinopathy can be predicted using machine learning. Training two validation datasets using deep learning to detect and grade diabetic retinopathy and macular edema achieved a high specificity and sensitivity for detecting moderately severe retinopathy and macular edema after each image was graded by ophthalmologists between three and seven times
6. Thomas Schaffter	Evaluation of combined artificial intelligence and radiologist assessment to interpret screening mammograms	To evaluate whether AI can overcome human mammography interpretation limitations	An AI algorithm combined with the single radiologist assessment was associated with higher overall mammography interpretive accuracy in independent screening programs compared with a single radiologist interpretation alone
7. Diego A. Hipolito Canario	Using artificial intelligence to risk stratify COVID-19 patients based on chest x-ray findings	This study evaluated M-qXR's ability to serve as a risk stratification tool to help evaluate patients with possible COVID-19 diagnosis	M-qXR was found to have comparable accuracy in detecting radiographic abnormalities on CXR suggestive of COVID-19 when compared to radiological ground truth. The M-qXR algorithm has the potential to provide benefits in guiding the medical management of patients suspected of having

Continued

A. Background of digital transformation in health

References	Article title	Key contribution	Impact on our research model
			COVID-19 who present with a high likelihood of disease, and where timely viral testing is not feasible due to limited resources
8. Pradnya Brijmohan Bhattad	Artificial intelligence in modern medicine–the evolving necessity of the present and role in transforming the future of medical care	Aims to identify how artificial intelligence-based technology is reforming the art of medicine	AI has the potential to further improve patient care due to its ability to interpret more detailed and comprehensive data and as such, AI in medicine should be welcomed Incorporating artificial intelligence in our clinics has the potential to enhance physician efficiency. AI does not have the potential to replace the physician's expertise in the clinical presentation, which will remain the vital aspect of medical care Physicians will need to learn on appropriate applications and interpretation of artificial intelligence models in the new age of artificial intelligence. Artificial intelligence has its own limitations and cannot replace a bedside clinician
9. Kristen Wong	Perceptions of Canadian radiation oncologists, radiation physicists, radiation therapists, and radiation trainees about the impact of artificial intelligence in radiation oncology (national survey)	Investigate the perception of radiation oncologists, radiation therapists, medical physicists, and radiation trainees about AI and how AI will affect radiation oncology as a specialty	Radiation oncology professionals believe AI will be an important part of patient treatment in their future practices The fear about AI may be mitigated with further education programs about AI, which can lead to more confidence in the acceptance of AI
10. Samer Ellahham	Artificial intelligence: the future for diabetes care		Artificial intelligence will cause a paradigm shift in diabetic management through data-driven precision care. AI has changed the way diabetes is prevented, detected, and managed, which can help in bringing down the global prevalence of 8.8%

A. Background of digital transformation in health

2. Materials and methods

Study design and setting: An observational descriptive cross-sectional mapping study conducted in Saudi Arabia, over 1 month from June 6 to July 6, 2021.

Study population and sampling technique: All healthcare practitioners residing in all regions of Saudi Arabia and agreeing to participate in the study were eligible to be recruited for the study. The minimum sample size was calculated considering a level of confidence of 95%, expected prevalence of 50%, and precision of 0.05; it was found to be 420. To comply with the physical distancing rules in response to the COVID-19 pandemic, the recruitment of study participants was done through online invitations on different platforms of social media.

Study tool: The data were obtained using a predesigned online questionnaire sheet that has been validated and tested via Crombach testing. Additionally, close-ended preset questions were obtained from the article titled "Physician confidence in AI: An online mobile survey." Permission for the use of survey questions was obtained from Kwon Soon Hyo. The survey was in English. For every participant, the following data were collected: (a) Demographics (age, gender, profession, specialty, workplace location, institution type); (b) Knowledge ("AI is an area of computer science that stimulates human intelligence," "AI can understand and learn from experience and make decisions," "Neural network is a form of AI," "I know the difference between strong AI and weak AI," "Features (such as) product recommendation/spell correction use AI," "Speech recognition is a form of AI," "an electronic health record system is an AI," "Tabaud is a form of AI)"; (c) Perception (questions have been asked to analyze healthcare practitioners perception toward the future impact of AI); and (d) Application (questions have been asked to assess healthcare practitioner predictions toward the future use of AI).

Ethical considerations: Approval from the Ethics and Scientific Committees of the Saudi Commission for Health Specialties was obtained before initiation of the study. Informed consent was obtained online from each participant; the aim of the study was clearly explained and selecting the "agree to participate" icon was required before proceeding to the questionnaire items. Data were collected anonymously, and the confidentiality of collected data was guaranteed.

Data analysis: The collected data were statistically analyzed using the Statistical Package for Social Sciences (SPSS) software, version 23 (IBM, Chicago, Armonk, NY). Categorical variables were presented as numbers and percentages. Numerical variables were presented as means ± standard deviations. Testing for statistically significant differences between the subgroups was carried out. Comparing mean scores between two groups and more than two groups was carried out using student's t-test and analysis of variance (ANOVA), respectively. The level of significance was set at $P < .05$.

3. Results, materials, and methods

A total of 1154 participants (all healthcare practitioners) were included, with 68.4% males and 33.1% females. Participants were from both government (65.50%) and private (34.50%) institutions. A total of five professions and 17 different specialists participated from 12 regions of Saudi Arabia (Figs. 1–3).

A. Background of digital transformation in health

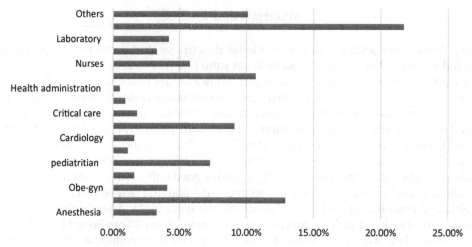

FIG. 1 Demographics: Specialties.

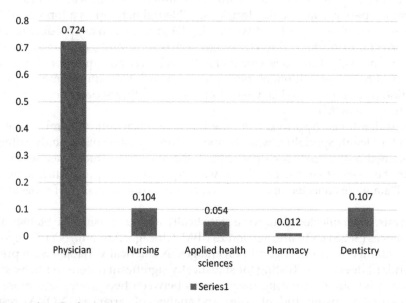

FIG. 2 Demographics: Professions.

To achieve our goals, we prepared a questionnaire that consisted of four main domains: demographics, knowledge, perception, and application Table 1 shows participant's overall knowledge about artificial intelligence.

Table 2 shows our questionnaire on perception. We asked our participants to rate their agreement from a scale of 1 to 10.

After collecting the responses, we primarily focused on profession-wise and location-wise responses.

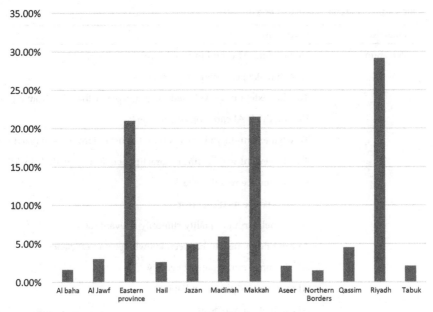

FIG. 3 Demographics: Regions.

TABLE 1 Questionnaire items related to knowledge.

Rank	Domain	Survey questions
1.	Knowledge	Do you think AI is an area of computer science that stimulates human intelligence?
2.		Do you think AI can understand and learn from experience?
3.		Do you think a neural network is a form of AI?
4.		Do you know the difference between strong and weak AI?
5.		Do you think features such as product recommendation and spell correction use AI?
6.		Do you think an electronic health record system uses AI?
7.		Do you think speech recognition is a form of AI?
8.		Do you think applications such as Tawakkalna and Tabaud use AI?
9.		AI has no emotional exhaustion or physical limitation
10.		From a scale of 1 to 10, rate your knowledge about AI

Table 3 demonstrates the percentage of responses (variable = profession) where approximately 81.5% agreed with the statement that AI in the future is going to be useful. Because this represents the whole population, we manually calculated the overall agreement.

In Fig. 4, an overall response to the first question in the perception domain has been shown.

A. Background of digital transformation in health

TABLE 2 Questionnaire items related to perception.

Rank	Domain	Questions
1.	Perception	Do you find AI useful in healthcare?
2.		AI will make healthcare affordable
3.		To what extent is an AI-based diagnosis better than one from a practitioner?
4.		Do you think AI can replace your job?
5.		To what extent do you trust A-based nonstandard, personalized treatment?
6.		To what extent will healthcare practitioners lose their skills to AI?
7.		Will AI reduce repetitive tasks?
8.		AI will reduce medical errors
9.		AI can deliver high-quality clinically relevant data
10.		AI can bridge the gap between knowledge and efficiency
11.		AI can compromise patient privacy
12.		AI can increase physician liability
13.		AI can make a diagnosis
14.		AI can assist in deciding treatment
15.		AI can contribute in biopharmaceutical research
16.		AI can provide medical assistance in underserved areas
17.		AI can assist in developing social development programs
18.		AI can optimize real-time patient flow
19.		AI can predict cost vs. benefit of a treatment
20.		AI would eventually be an essential part of your field

TABLE 3 Responses.

Profession	Agreed (s.a.+a)	Disagree (neutral+d+s.d.)	P value
Physician	711/835, 85.149%	14.851%	.000
Nurse	76/120, 63%	37%	
Applied health science	44/62, 70.967%	29.033%	
Pharmacy	9/14, 64.285%	35.7155	
Dentistry	100/123, 81.30%	18.7%	
Total	940 (81.455%)		

A. Background of digital transformation in health

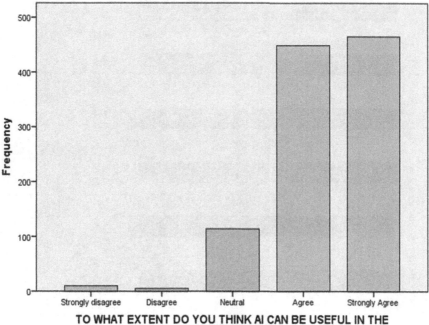

FIG. 4 AI is useful in the healthcare domain.

Fig. 5 shows responses throughout the country across regions. These are a visual representation of which regions are likely to accept AI better.

3.1 AI might compromise patient privacy

According to our null hypothesis, we assumed that healthcare providers from different domains believe that AI will not harm patient privacy. According to the result in Table 4, the P value is insignificant ($P = .53 > .05$). This fails to reject our null hypothesis.

The bar graph in Fig. 6 is confirming respondents' attitude that AI can compromise patients' privacy and access to critical health information?

3.2 AI might increase physician liability

According to our null hypothesis, the belief is that physician liability while using AI-based healthcare will increase. The data seen in Table 5 denote an insignificant value ($P = .201 > .05$). This fails to reject our null hypothesis. (See Table 6.)

A. Background of digital transformation in health

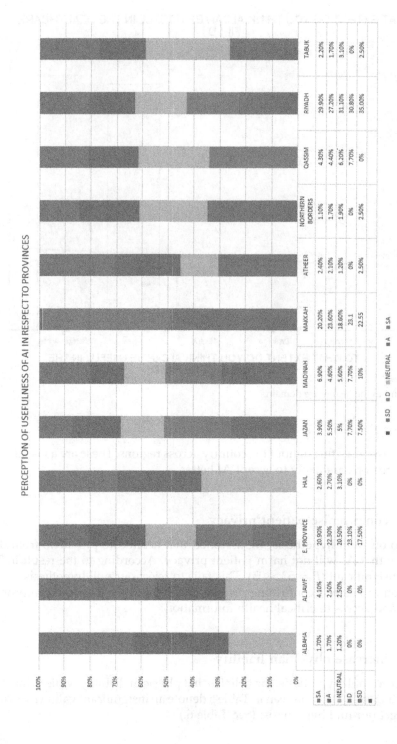

FIG. 5 Perception of the usefulness of AI (per province in Saudi Arabia).

TABLE 4 AI might compromise patient privacy.

Profession	Agree	Disagree	P
Physician	360/835, 43.113%	56.886%	.533
Nurse	51/120, 42.500%	57.5%	
Applied health science	28/62, 45.161%	54.838%	
Pharmacy	9/14, 64.285%	35.714%	
Dentistry	55/123, 44.715%	55.285%	

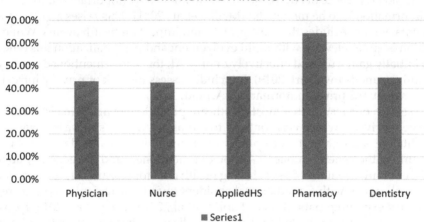

FIG. 6 AI might compromise patient privacy.

TABLE 5 AI might increase physician liability.

Profession	Agree	Disagree	P
Physician	364/835, 43.592%	56.407%	.201
Nurse	37/120, 30.833%	69.166%	
Applied health science	33/62, 53.225%	46.774%	
Pharmacy	5/40, 35.714%	64.285%	
Dentistry	53/123, 43.089%	56.910%	

TABLE 6 Usage of AI in respective workplace.

Profession	Agree	Disagree	P
Physician	215/835, 25.748%	74.251%	.017
Nurse	38/120, 31.667%	68.33%	
Applied health science	25/62, 40.322%	59.677%	
Pharmacy	2/14, 14.285%	85.714%	
Dentistry	29/123, 23.577%	76.422%	

A. Background of digital transformation in health

According to our null hypothesis, AI is being vastly used in all the provinces of Saudi Arabia, regardless of the advancement of technology in general. Our result gives a significant value ($P = .017 < .05$). This rejects our null hypothesis.

4. Discussion

4.1 AI and patient privacy

With the rapid emergence of AI, we have come to know how important our data is and how likely our information is to be monetized (Maliha et al., 2021). This raises a great concern for our health data, which includes the most private and important health events. While privacy of information is quite relevant, with rapid evolvement and personalization this criterion can be met. To initially gain trust and effectively deploy AI, the below-mentioned roadmap is inevitable (Sullivan and Schweikart, 2019). It includes assess the risk of privacy intrusions and initiatives to safeguard private information (Arnold, 2021).

We have seen in our results that most healthcare practitioners agree that AI will not compromise patient privacy, which was contrary to our anticipated response but was consistent with a qualitative study about perceptions of using public health data for AI research conducted among the general public.[a] Although practitioners and the general public on average are in harmony about the idea of sharing health data for research, a study conducted by Paprica and colleagues with Canadian stakeholders found participant suspicion regarding private industry exploiting patient data (Paprica et al., 2019; Teng et al., 2019; Ipsos MORI, 2016). Similarly, a study conducted by Kim and colleagues has shown patient hesitance to share health data with private for-profit organizations (Royal Society, n.d.).

4.2 AI and physician liability (our paper versus others)

Liability in healthcare practice denotes taking responsibility and paying penalties to consumers when malpractice is proven. Current tort law (a tort is a civil claim requested by the plaintiff against a defendant causing a harmful act on the former) includes certain aspects such as physician liability, a healthcare organization's liability, manufacturer liability, pharmaceutical liability, etc. (Bathaee, n.d.).

In our study, it is evident that all except practitioners from applied health sciences disagree with the statement that "AI will increase (the) practitioner's liability." It's interesting to know that these liabilities aren't applicable to AI. According to Yavar Bathaee, "The law is built on legal doctrines that are focused on human conduct, which when applied to AI, may not function" (ResearchGate, n.d.). Mathew Scherer said that black box AIs are unpredictable in nature. If it is harmful after being released into the world, who is to blame? Designers or AIs? Because even according to current law, physician intervenes between manufacturers and consumer, thus preventing plaintiff directly suing the manufacturers (Allain, n.d; Bathaee, n.d). Now this raises questions, that being in frontline, will application of AI for oversimplification of healthcare system to have a unified service will make practitioner's the prime plaintiff or not. If it does, circumstances where "using the software had been the only option" would be an exception or not? And how would we finally ensure patient's right in case the prime plaintiff is not a human.

4.3 AI and usage in workplace

Regarding whether AI is being used in the workplace, our participants have significantly refused using AI-based software.

This brings us to the question of whether participants are aware of electronic health record (EHR) systems using AI. According to a study conducted by Sana A. AlSadrah, an EHR system was introduced in Saudi Arabia in 2008. However, its widespread application has been delayed due to a lack of interest in understanding the benefits of an EHR system as well as inadequate communication between frontline doctors and IT experts. Nowadays, except for a few rural hospitals and primary healthcare centers, most everyone in Saudi Arabia is using AI (AlSadrah, 2020).

4.4 AI and current applications

There are various specialties where AI techniques/functional applications play a promising role such as in medical imaging, public health, genetics, drug discovery, and medical informatics (Bhattad and Jain, 2020).

A 2020 article published by Samer Ellahham shows that "deep learning algorithms have been developed to automate the diagnosis of diabetic retinopathy" (Lytras and Chui, 2019). The AI-based screening of the retina is a feasible, accurate, and well-accepted method for the detection and monitoring of diabetic retinopathy. A high sensitivity and specificity of 92.3% and 93.7%, respectively, have been reported for automated screening of the retina. Patient satisfaction for automated screening is also high with 96% of patients reported as being satisfied or very satisfied with this method (Ellahham, 2020). Mintz and Brodie talked about how widely used AI is in healthcare and in different specialties such as cardiology. "The application of machine learning (ML) and AI results in faster interpretation and diagnosis in many areas of cardiology. Electrocardiogram readings are automatically interpreted, and echocardiography with three-dimensional (3D) mode cardiac imaging automatically provides measurements of cardiac function" (Mintz and Brodie, 2019) (Fig. 7).

Regarding oncology, AI has a role in the detection of lung cancer. AI algorithms have been shown to be more effective than a human. In a study using 2186 stained histopathology whole-slide images of lung adenocarcinoma and squamous cell carcinoma, Yu et al. demonstrated the accuracy of AI for pathological diagnosis.

Their results support that AI can anticipate patient prognosis, which can help enhance patient care via the determination of oncological treatment (Mintz and Brodie, 2019).

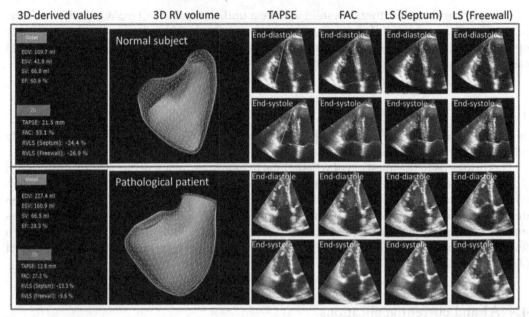

FIG. 7 Example of three-dimensional echocardiographic evaluation of right ventricular (RV) volumes and function in a normal subject (upper panels) and in a pathological patient (lower panels) with the onboard dedicated software (Tamborini et al., 2019).

References

Ahuja, A.S., 2019. The impact of artificial intelligence in medicine on the future role of the physician. PeerJ 7 (e7702), e7702. (Internet, cited 2022 Jan 10). Available from: https://www.ncbi.nlm.nih.gov/labs/pmc/articles/PMC6779111/.

Allain, J.S. (n.d.). From Jeopardy! To Jaundice: The Medical Liability Implications of from Jeopardy! To Jaundice: The Medical Liability Implications of Dr. Watson and Other Artificial Intelligence Systems Dr. Watson and Other Artificial Intelligence Systems. Lsu.Edu. Available from: https://digitalcommons.law.lsu.edu/cgi/viewcontent.cgi?referer=https://www.google.com/&httpsredir=1&article=6423&context (Retrieved 10 January 2022).

AlSadrah, S.A., 2020. Electronic medical records and health care promotion in Saudi Arabia: an overview. Saudi Med. J. 41 (6), 583–589. https://doi.org/10.15537/smj.2020.6.25115.

Arnold, M.H., 2021. Teasing out artificial intelligence in medicine: an ethical critique of artificial intelligence and machine learning in medicine. J. Bioethical Inq. 18 (1), 121–139. https://doi.org/10.1007/s11673-020-10080-1.

Bathaee, Y. (n.d.). The artificial intelligence black box and the failure of intent and causation. Harvard Edu. Available from: https://jolt.law.harvard.edu/assets/articlePDFs/v31/The-Artificial-Intelligence-Black-Box-and-the-Failure-of-Intent-and-Causation-Yavar-Bathaee.pdf (Retrieved 10 January 2022).

Bhattad, P.B., Jain, V., 2020. Artificial intelligence in modern medicine—the evolving necessity of the present and role in transforming the future of medical care. Cureus (Internet) 12 (5), e8041. https://doi.org/10.7759/cureus.8041.

Bohr, A., Memarzadeh, K., 2020. The rise of artificial intelligence in healthcare applications. In: Artificial Intelligence in Healthcare. Elsevier, pp. 25–60.

Ellahham, S., 2020. Artificial intelligence: the future for diabetes care. Am. J. Med. (Internet) 133 (8), 895–900. https://doi.org/10.1016/j.amjmed.2020.03.033.

Ipsos MORI, 2016. The One-Way Mirror: Public Attitudes to Commercial Access to Health Data. Ipsos MORI, London UK. Available: https://www.ipsos.com/sites/default/files/publication/5200-03/sri-wellcome-trust-commercial-access-to-health-data.pdf.

Lytras, M.D., Chui, K.T., 2019. The recent development of artificial intelligence for smart and sustainable energy systems and applications. Energies 12 (16). https://doi.org/10.3390/en12163108. Art. no. 3108.

Maliha, G., Gerke, S., Cohen, I.G., Parikh, R.B., 2021. Artificial intelligence and liability in medicine: balancing safety and innovation. Milbank Q (Internet) 99 (3), 629–647. Available from: https://www.milbank.org/quarterly/articles/artificial-intelligence-and-liability-in-medicine-balancing-safety-and-innovation/. (cited 2022 January 10).

Mintz, Y., Brodie, R., 2019. Introduction to artificial intelligence in medicine. Minim. Invasive Ther. Allied Technol. (Internet) 28 (2), 73–81. https://doi.org/10.1080/13645706.2019.1575882.

Paprica, P.A., de Melo, M.N., Schull, M.J., 2019. Social licence and the general public's attitudes toward research based on linked administrative health data: a qualitative study. CMAJ Open 7, E40–E46.

ResearchGate. (n.d.). Available from: https://www.researchgate.net/deref/https%3A%2F%2Fwww.nytimes.com%2F2018%2F01%2F25%2Fopinion%2Fartificial-intelligence-black-box.html (Retrieved 10 January 2022).

Royal Society, Public Views of Machine Learning: Findings From Public Research and Engagement Conducted on Behalf of the Royal Society. Available from: https://royalsociety.org/~/media/policy/projects/machine-learning/publications/public-views-of-machine-learning-ipsos-mori.pdf.

Sullivan, H.R., Schweikart, S.J., 2019. Are current tort liability doctrines adequate for addressing injury caused by AI? AMA J. Ethics 21 (2), E160–E166. https://doi.org/10.1001/amajethics.2019.160.

Tamborini, G., Cefalù, C., Celeste, F., Fusini, L., Garlaschè, A., Muratori, M., et al., 2019. Multi-parametric "on board" evaluation of right ventricular function using three-dimensional echocardiography: feasibility and comparison to traditional two-and three dimensional echocardiographic measurements. Int. J. Cardiovasc. Imaging (Internet) 35 (2), 275–284. https://doi.org/10.1007/s10554-018-1496-9.

Teng, J., Bentley, C., Burgess, M.M., et al., 2019. Sharing linked data sets for research: results from a deliberative public engagement event in British Columbia, Canada. Int. J. Popul. Data Sci., 4.

YESSER E-Government Program, n.d. الرعاية الصحية (Internet) Gov.sa. Available from: https://www.my.gov.sa/wps/portal/snp/aboutksa/HealthCareInKSA (cited 2022 January 10).

A. Background of digital transformation in health

Enabling technologies for digital transformation (an example-oriented approach)

Internet of things (IoT) and big data analytics (BDA) in healthcare

Prableen Kaur

Department of Mathematics, Chandigarh University, Chandigarh, Punjab, India

1. Introduction

Data are being generated at an ever-increasing rate from many disciplines such as finance, business, healthcare, and so on. This new data in the healthcare industry comprises patient electronic health records (EHRs), lab results, database management, medical imaging, and patient monitoring, among other things. A decade ago, all this information was saved in hard copy format and was thrown away after a certain time period because it was considered a waste; now, these data are preserved in digital form to extract as much value as possible from them. This digitization occurred as a result of large volumes of data (also known as "big data") acquired from the medical industry being identified with promising and supporting suggestions for a broader range of healthcare functions as well as decision support, disease observation, and population health management. Furthermore, the amount of information generated is overwhelming for the healthcare business due to the five Vs of big data: volume, velocity, variety, veracity, and value (Raghupathi and Raghupathi, 2014).

Big data in the healthcare is described as massive volumes of information or data created by the conversion because it helps in assembling patient records and in managing the performance of hospitals. Also, these data are massive and complicated for traditional technologies. Analysis of these data uses new researched technologies such as big data analytics (BDA) and the Internet of Things (IoT) to make value from the dataset. These technologies play very important roles in the digital transformation of healthcare by using advanced tools and techniques for collecting personal health data from innumerable consumers in an increasing trend.

IoT and BDA are two different terms, but they have a common goal and work together to advance in different areas. The main purpose of these two techniques is to analyze the data.

The IoT generates data from a variety of sources, requiring BDA tools to analyze it. Together, these two technologies will work to provide valuable information.

The IoT and BDA have been of tremendous help in the current pandemic situation where every country is battling COVID-19. During the COVID-19 pandemic, both these technologies worked to alter the healthcare system. The data provided by the IoT device are just as useful as the analytics that can be derived from it. Big data will drive the shift from hypothesis-based research to data-driven research. On the other hand, the IoT helps monitor and analyze different levels of communication between different sensor signals and traditional big data. This will enable new remote diagnostic tools along with a better understanding of the disease, leading to innovative healthcare solutions (Lirette, 2019; Digiteum Team, 2019).

1.1 The IoT

The world has never been the same after the introduction of the Internet. The Internet has helped to link everything in the world while also providing several advantages. The IoT is a framework of networked computer systems as well as digital and mechanical devices capable of exchanging data over a predefined network without any human interaction at any level. In a nutshell, it can link any gadget to the Internet and allow us to turn it on and off as well as connect it to other devices. The "thing" in IoT refers to a human or any equipment that has an IP address given to it. Using embedded technology, a thing captures and sends data via the Internet without the need for human involvement. It assists people in making decisions by allowing them to engage with the external world or internal states (Malhotra, 2018; Singh, n.d.).

People can use apps or web applications to remotely operate this network and gain real-time insights into its operation. Furthermore, any linked device may be outfitted with sensors that generate data on its status, functionality, and surroundings, and this data can be continuously sent to the backend system for analysis. This is a high-level concept of how a linked area may operate. It comprises home automation systems, cell phones, and wearable gadgets, among other things. When we think about the IoT, the next phrase that comes to mind is "ubiquitous networking." Essentially, it is the technical definition of the Internet, which is available at all times (Singh, n.d.; Bhatt and Bhatt, 2017).

1.2 Big data

Telescopes, sensor networks, high-throughput devices, and streaming machines, for example, are rapidly improving data production sources, and these settings create vast volumes of data. Big data refers to a collection of vast volumes of data that might be organized, unstructured, or semistructured (Manogaran et al., 2017; Rghioui et al., 2020).

As seen in Fig. 1, big data can be segmented into the five Vs: volume, velocity, variety, veracity, and value

1. Volume: The large volume does, in fact, reflect a large amount of data. Gradually, data creation sources have been supplemented, resulting in a wide range of data types such as text, video, audio, and large-scale photographs. To process the massive amounts of data, our traditional data processing systems and procedures must be improved. This refers to

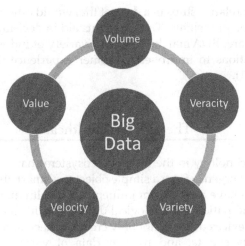

FIG. 1   The five Vs of big data.

the massive quantity of data generated by hospitals via applications, portals, websites, and electronic health records (EHRs).

2. Veracity: This data veracity truly reflects big data. The term "veracity" refers to the capacity to comprehend data rather than the quality of the data. It is critical that the organization execute data processing to avoid "dirty data" from collecting in the systems.

3. Variety: The diversity of the data does, in fact, reflect huge data. The number of data formats available today is likewise growing. Most businesses, for example, employ data formats such as databases, Microsoft Excel, and comma-separated values (CSV), which may all be saved in a plain text file. Nonetheless, the data may not always be in the expected format, causing processing challenges. To overcome this problem, the organization must establish a data storage system capable of analyzing a wide range of data.

4. Velocity: The rate of incoming data has skyrocketed; this velocity really symbolizes big data. The term "velocity" refers to the pace at which data are generated. The data explosion caused by social media has transformed and caused data diversity. People today are less bothered with previous posts (tweets, status updates, and so on) and are more interested in new information.

3. Value: Big data is represented by the value of data. Having constant volumes of data is useless unless it can be converted into value. It is critical to recognize that the presence of huge data does not always imply that it is valuable. When undertaking BDA, the advantages and expenses of analyzing and acquiring huge datasets are more essential.

1.3 BDA

The difficult aspect is analyzing massive datasets because typical data management solutions do not operate well. BDA is the strategy necessary to address this difficult portion.

It seeks to find patterns or relationships in a dataset that would otherwise be inconspicuous to find important information or insights. These insights aid in decision-making abilities, fraud prevention, and much more. BDA may be used in a variety of industries ranging from education to telecommunications to improved customer experience (Manogaran et al., 2017; TestingXperts Team, 2021).

2. The IoT and healthcare

IoT is a developing technology in the Internet ecosystem that combines real-time linked things. Because of the convergence from a simple object to a smart object, it is useful in a wide range of sectors. This will have a long-term influence on health monitoring, administration, and clinical care related to patient physiological data. In healthcare, the IoT refers to a network of linked medical devices that can not only create, collect, and store data, but also connect to a network, analyze data, and transfer data of various types such as medical photographs, physiological and vital body signs, and genetic data. In simplified terms, sensor devices may collect many types of patient data and receive feedback from healthcare providers. Continuous glucose monitoring for insulin pens is an example of IoT healthcare that works well for diabetics (Ananda Kumar and Mahesh, 2021).

All sensor devices can interact with one another and, in some situations, perform critical steps that might give immediate assistance or even save a life. An IoT healthcare gadget, for example, may make intelligent judgments such as notifying a healthcare center if an elderly person falls down. After collecting idle data, an IoT healthcare device would communicate this important information to the cloud so that clinicians could act on it—assess the overall patient state, determine whether an ambulance is needed, determine what sort of assistance is needed, and so on. Following are few applications of IoT (Inspirisys Solutions Limited, 2021; Singh et al., 2020) (Table 1):

3. BDA and healthcare

BDA is a new paradigm that is primarily intended to analyze, manage, and accurately extract usable information from vast quantities of datasets that are extremely similar to a specific patient in a very short period of time. Additionally, this new technology-based analytics strategy completely modifies the healthcare sector by allowing the correct decision to be made for the right patient at the right time. Meaningful use and compensation for performance are emerging as key new variables in today's healthcare environment as healthcare models change. Although profit is not and should not be the primary motive, healthcare organizations must acquire the available tools, infrastructure, and methodologies to properly exploit big data or risk losing possibly millions of dollars in revenue and profits. Big data contains features such as diversity, velocity, and, in the context of healthcare, veracity. Available analytical approaches may be used on huge amounts of existing (but presently unanalyzed) patient-related health and medical data to gain a better knowledge of outcomes, which can then be implemented at the point of treatment. Individual and population data, ideally,

TABLE 1   Overview of Internet of things (IoT) enabled applications.

1	Remote temperature monitoring for vaccines	IoT aids in vaccination management by tracking vaccine temperature and humidity and warning when they exceed the prescribed threshold.
2	Telehealth Consultations	People can use video chats to consult with a doctor from the comfort of their own homes for nonemergency medical issues. This will minimize the number of consultation trips to the doctor's office.
3	Blood coagulation devices	IoT connectivity aids in self-monitoring for blood coagulation and wirelessly transmitting findings to healthcare practitioners.
4	Connected inhalers	The most common inhalers may be converted into smart linked inhalers using artificial intelligence (AI)-enabled IoT sensors.
5	Remote monitoring	Doctors may treat patients with IoT devices that monitor glucose levels, heart rate, blood pressure, and other parameters. In the event of an emergency, these people can be swiftly hospitalized.
6	Digital diagnostics	Kinsa, a smart thermometer, can detect fever spikes caused by viruses across the United States and notify municipalities about the trend.
7	Robot assistance	To decrease contamination from COVID-19, IoT robots are being used in hospitals to disinfect places, clean patient rooms, and administer medications.
8	Emergency services	In an emergency, IoT can immediately calculate the distance and arrange to transport patients to the nearest hospital for treatment.
9	Medicine reminder	The IoT can remind patients to take their medications on time to avoid missing a dose.
10	Air quality sensors	IoT devices improve the safety of surgical patients by providing dashboards on hygiene and safety issues in the operating room.
11	Depression Monitoring	Indications of a panic attack can be identified using a paired wristband.
12	Connect all medical tools and devices through the Internet	During COVID-19 treatment, the IoT connected all medical instruments and devices through the Internet, delivering real-time data.
13	Accurate forecasting of virus	Based on the given data, the application of specific statistical approaches can also aid in forecasting the scenario in the future. It would also assist the government, physicians, professors, and others in planning for a better working environment.
14	Internet-connected hospital	The use of IoT to support pandemics such as COVID-19 necessitates a fully integrated network within hospital grounds.

would advise each physician and patient during the decision-making process, assisting in determining the most effective treatment plan for that specific patient (Raghupathi and Raghupathi, 2014; Haleem et al., 2020).

Because of its capacity to store enormous amounts of data, big data can save the whole medical history, travel history, and fever symptoms of all patients. By supplying the acquired data, this technology aids in the identification of infected patients and the subsequent study of the level of danger. It also aids in identifying those who may have been in touch with a virus-infected patient as well as suspicious instances and other disinformation with the relevant data. It aids in the early identification of infected patients as well as the analysis and identification of those who may get infected by this virus in the future. It also tracks and monitors people's movements and overall health management. Following that, it aids in the rapid creation of new medications and medical equipment for present and future medical demands. It utilizes previously inhabited viral data and supports significant new insights for pandemic.

The use of big data technology in healthcare has a great variety of good consequences, including life-saving phenomena. Nowadays, all healthcare organizations gather and store all patient data and information. This sort of innovative technology is employed in healthcare to achieve better findings that can benefit diagnosis and disposition for patients. EHRs, machine-generated/sensor data, health information exchanges, patient registries, portals, genetic databases, and public records are some of the data types used in healthcare applications. Public records are significant sources of big data in healthcare, necessitating effective data analytics to address the associated healthcare issues. These large amounts of health data are kept in a medical server (MS), clinical database (CDB), and other clinical data repositories (CDR) for further study. Storage infrastructures are generally used to store, process, analyze, manage, and retrieve massive volumes of data to facilitate people's lives. As a result, it not only provides information to comprehend symptoms, illnesses, and treatments, but it also alerts and predicts outcomes at early stages to make the appropriate judgments (Kumar and Singh, 2019; Biswas, 2022).

4. A further description of pandemic use

Investigating the possibility of COVID-19 viral transmission through indirect contact, medical professionals may remotely monitor COVID-19 patients and self-quarantine people thanks to the smooth connectivity offered by IoT applications/devices. This cutting-edge technology ensures that all coronavirus-infected people are isolated. It is beneficial to have a proper monitoring system in place during quarantine. Using the Internet-based network, all high-risk patients may be readily followed. The medical personnel statistics are not accessible in comparison to the overall number of affected or suspected COVID-19 viruses. With the data security provided by IoT apps, medical professionals may gather all the relevant parameters for these patients in one location and decide on the next course of action. Because IoT apps are simple to use, patients may handle them on their own. The possibility of COVID-19 viral transmission via indirect contact is being investigated. Because of the fluid communication provided by IoT applications/devices, medical practitioners may remotely monitor COVID-19 patients and self-quarantine people. This cutting-edge technology ensures that all patients infected with the coronavirus are isolated. Having a proper monitoring system in place during quarantine is important. All high-risk patients can be easily tracked via an internet-based network. Medical staff figures are not available in comparison to the total number of COVID-19 viruses afflicted or suspected. Medical practitioners may gather all of the important characteristics for these patients in one area and decide on the next course of action thanks to the data security provided by IoT apps. Patients may be able to handle IoT apps on their own because they are simple to use. Big data contact-tracking analytics may aid with disease outbreaks by using many data sources, such as social media postings with metadata and tags, passenger lists, metro smartcards, vehicle records, and credit card use, all of which are useful sources of data. Although not exact, evaluating attributes of relevance from location metadata/tags on social media to confirm that a person was at a specific area at a certain time might create a tracking model that could trace individuals, even if they do not have tracking equipment or mobile phones with them (Wang et al., 2021).

The IoT and BDA work together to anticipate pandemics with neural networks and compare the findings to other machine learning techniques. They also allow descriptive, predeictive and prescriptive analytics for meaningful healthcare decisions. They also concentrate on enhancing the performance and quality of auxiliary power unit health monitoring services, with direct application in aviation and aerospace. They also assess COVID-19's influence on mental health and well-being. Designing mitigation techniques and allocating resources based on illness incidence rate prediction can serve critical healthcare decision making and optimization.

5. Use case

5.1 Aarogya Setu App

The Aarogya Setu App is a COVID-19 (coronavirus) mobile tracking application created by the National Informatics Centre (NIC), which is part of India's Ministry of Electronics and Information Technology. The Indian Prime Minister has urged Indian citizens to download this app on their mobile devices. This app is based on physical distance, and Bluetooth technology is used to detect critical information, including the counting schedule, proximity, location, and time scale. The app was a government endeavor to track and trace infected persons. It operates on the physical closeness concept. The app is a terrific effort at this time, and it has assisted every resident. It does have certain limitations.

The program is developed in such a manner that it notifies users whenever they come in contact with a COVID-positive individual. The tracking is done via Bluetooth technology and location-generated social graphs or global positioning system (GPS), which indicates the user's interaction with anybody who has tested positive and alerts them. With the use of GPS and Bluetooth sensors, it identifies and monitors the user's movement. Using its database and algorithms, it will also detect other nearby cell phones that have the program loaded and will send a signal if they come in contact with infected people. The app also includes directions on how to self-isolate and what to do if somebody gets symptoms. The app suggests that Bluetooth and location services be turned on at all times (Gupta et al., 2020).

Using IoT technologies such as Bluetooth and GPS sensors, this program has assisted in tracking the number of persons who have come into contact with recently afflicted individuals. Furthermore, all these data are maintained in their database, which is not open source, and big data technologies aid in the detection of infected instances and additional risk assessments. In simple terms, if one of the users tests positive, the app warns the other user, which aids the authorities in tracking future COVID-19 instances. Along with these warnings, the program includes advice for any viral symptoms discovered in a neighborhood. Aside from

Bluetooth and location data, the app gathers demographic information from users such as gender, phone number, age, complete name, and international travel history. It also asks the user whether they work in the provision of important services and if they would be prepared to assist in times of need. Following that, the government may use these data to track down possible instances and find volunteers willing to help in times of disaster (Gupta et al., 2020).

5.2 Remote patient monitoring (RPM)

Residents in many regions of the world live miles away from the nearest hospital. As a result, it takes time for them to reach healthcare facilities in the event of an emergency. Similarly, it becomes difficult for healthcare practitioners to see patients with chronic diseases on a regular basis. Remote patient care enabled by IoT can tackle the problem of time-consuming commutes (Kajeet Team, 2021).

Today's healthcare systems can leverage data obtained from patients via smartphones or different forms of smart body sensors to perform routine testing at home. The whole of a patient's data will now be captured in each patient's EHR, which contains the patient's health information in electronic format. The saved data are then used by BDA to offer a clear analysis of the patient's current health state. With the increase of chronic diseases in both rural and urban locations, IoT connection has aided in the delivery of timely and cost-effective RPM treatments regardless of patient location or doctor availability. This method of administration has been shown to be very beneficial in rural areas where skilled medical staff are frequently absent. Physical visits to the doctor's office for normal consultations are frequently impossible during the present COVID-19 outbreak.

This gives hospitals, patients, insurers, and even government organizations a number of incredible benefits, including (Fig. 2):

(a) Convenience

RPM makes it simpler for both healthcare providers and their patients to receive the treatment they require when they require it. Improved convenience has also resulted in increased efficiency in healthcare delivery. To demonstrate how this works, RPM has made remote care

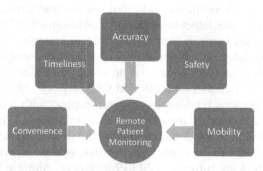

FIG. 2   Remote patient monitoring as a DT service.

simple. Because RPM systems provide near-instant communication capabilities, automated scheduling of virtual meetings has replaced in-person visits without hurting doctor-patient face time.

(b) Timeliness

Healthcare personnel may attend to their patients considerably more swiftly using RPM. RPM's continuous monitoring of a patient's vitals aids in faster reaction times, which saves lives, and eliminates delayed diagnoses, which can significantly affect recovery durations. It also expedites the commencement of important medications.

(c) Accuracy

The accuracy of RPM devices can increase efficiency and collaboration among patients, doctors, and those aiding in a patient's care, such as loved ones, external consultants, and subject matter experts. By bringing all these people together on a single platform and including other services such as GPS location, situations may be treated more swiftly and efficiently. Various capabilities may be implemented into the RPM system over time to enable insurers and government entities to process claims and offer services that patients require. These would be based on automated reports created by secure RPM devices that gather patient data, mix it with doctor suggestions, and communicate it to the appropriate insurance or government office, rather than long claim forms filled out by the patient.

(d) Safety

Physical and digital security may be implemented in your RPM system to guarantee that access to patient health information is only granted as needed and that EHRs and patient data comply with HIPAA regulations.

(e) Mobility

Mobile solutions for health services can enhance patient satisfaction with care. Smart health technologies such as Fitbit and other health-monitoring technologies have gained general acceptance, indicating that more people are accepting RPM as a viable option. Doctors may increase their participation in the treatment process and give material to patients in an easy-to-understand style by giving consumers the choice to manage their care remotely. This technique offers higher satisfaction by resolving patient complaints in real time at little or no expense.

The health sector has the potential to save tens of billions of dollars each year with the correct IoT infrastructure in place. According to the Federal Communications Commission's (FCC) National Broadband Plan, deploying RPM technologies in conjunction with EHRs may save the healthcare sector $700 billion over the next two decades.

6. Challenges

(a) Real-time BDA is a critical demand in healthcare. It is necessary to solve the time gap between data collecting and processing. However, big data presents a number of obstacles that, if overcome, might make life easier. It is simple to utter the word eliminate, but it is not so simple to put it into action. The list below highlights a few IoT and BDA issues that we are now facing (Chaudhari et al., 2020; Team, 2020).

(b) New technology: According to World Health Organization data, the age group most at risk is those over 65. People with heart disease and diabetes are also at significant risk in this pandemic condition caused by COVID-19. It is tough for these folks to manage IoT devices at their age. They require assistance in order to read correctly. This will be the most difficult hurdle for IoT.

(c) Skilled resource pool: To undertake big data analysis, a data scientist and a data analyst are required. The essential competence for BDA is already in limited supply.

(d) Power source: In this case, in industries such as electricity, only 30–40% of individuals are in charge of the power supply. Any type of power outage will take time to fix. In the event of a power outage, the Internet connection will be lost. As a result, it is possible that data will not be uploaded or disseminated in a timely manner.

(e) People who are uneducated: People in rural areas are not trained to use IoT devices. In India, more than 60% of the population lives in rural regions. As a result, we must teach these folks how to use these technologies.

(f) Massive produced data inputs: Thousands of devices in a single healthcare institution, plus a thousand more streaming data from faraway places—all in real time—will create massive volumes of data. The data generated by IoT in healthcare will almost certainly increase storage requirements from terabytes to petabytes. When used appropriately, AI-driven algorithms may be highly valuable. AI-driven algorithms and the cloud, when used correctly, may help make sense of and organize enormous amounts of data, but this method will take time to evolve. As a result, developing a large-scale IoT healthcare solution will take a significant amount of time and work.

(g) Concerns about data security and privacy: The security risks context will grow as a result of IoT devices. IoT healthcare provides tremendous benefits to the sector, but it also introduces significant security flaws. Hackers might get access to medical equipment connected to the Internet and steal or manipulate the data. They can also go a step further and attack IoT devices with notorious ransomware, infecting a whole hospital network. That implies that the hackers will take captive patients as well as their heart rate monitors, blood pressure readings, and brain scanners. Furthermore, the revelation of personal health information poses a significant danger. Existing policies must be reviewed to ensure that protected health information (PHI) data is handled with the greatest care.

(h) Funding sources: To provide improved treatment, funding patterns must be reconsidered. Incentives should be differentiated between physicians who offer high-quality treatment and those who fall short of expectations.

(i) Governance: Governance policies will be influenced by BDA. Existing legislation, governance, and information management processes would be severely disrupted.

(j) Outdated infrastructure: The existing software infrastructure is out of date. Many hospital IT systems are out of date. This will prevent appropriate IoT device integration. As a result, healthcare institutions will need to overhaul their IT procedures and implement newer, more current technologies. They will also need to use virtualization [technology such as software-defined networking (SDN) and network functions virtualization (NFV)] as well as ultrafast wireless and mobile networks such as advanced LTE or 5G.

(k) More research: BDA is still in its infancy, and present tools and approaches are incapable of addressing the difficulties connected with big data. Big data may be considered as large systems that provide enormous issues. As a result, extensive study in this subject will be necessary to address the difficulties confronting the healthcare system.

(l) Packaged: It is well acknowledged that there are several big data applications that can revolutionize our healthcare operations, increase the quality of care, and deliver personalized patient care. A BDA platform in healthcare must at the very least cover the essential functionalities required for data processing. Platform evaluation factors might include availability, continuity, simplicity of use, scalability, the capacity to manipulate at various levels of granularity, privacy and security enablement, and quality assurance. Furthermore, while the majority of platforms now accessible are open source, the standard benefits and restrictions of open-source platforms apply. To be successful, BDA in healthcare must be packaged in a menu-driven, user-friendly, and transparent manner.

7. Future

While the world is fighting COVID-19, numerous technologies, including IoT and BDA, have grabbed the attention of academics and have had a significant influence on our lives and the planet. Throughout the pandemic, IoT has demonstrated promising results in social, psychological, and economical elements of life. The significant issues presented by IoT and BDA constitute a future research worry for academics seeking to make the world safer and more automated. Because of IoT's massive development in every area, its adoption opens up hundreds of opportunities as well as long-term solutions to the world's real-time difficulties in education, health, business, and security (Nižetić et al., 2020; Undavia and Patel, 2020).

In the future, IoT and BDA will monitor a patient's vital signs in real time. This technology will digitally capture all detailed information to prevent recurring complications with the COVID-19 patient's therapy. The adoption of cutting-edge technology will significantly improve healthcare practice, and doctors will be required to use them. The IoT is a sophisticated evolving technology with several applications in providing accurate medical treatment, opening up an efficient means to analyze important data, information, and testing. The future includes applications for managing inventories in the medical profession and the medical supply chain to ensure that the correct item is available at the right time and location. Intelligent IoT gadgets would function autonomously. There will be data storage via private and public clouds as well as software, allowing illness detection and follow up to be more efficient. This revolutionary information system breakthrough will enable intelligent healthcare services in the medical environment(Singh et al., 2020).

8. Conclusion

The IoT and BDA look to be major parts of the healthcare system's transition today. The advancement of IoT device technology has the potential to significantly enhance not only a patient's health and assistance in urgent situations, but also the productivity of health professionals and hospital processes. To combat COVID-19, IoT provides a smart dedicated gateway. IoT does this by using its robust integrated network. Healthcare monitoring equipment is linked to the Internet as a result of this significant interconnectivity. All data are remotely monitored at the healthcare center. This revolution is predicted to help healthcare services in a variety of sectors ranging from clinical treatment to biological information and general management. The Aarogya Setu application and remote patient management make good use of IoT and BDA technology while also bringing exciting innovations to the worlds of health services and care giving.

However, some obstacles remain, and more study is needed for the future of healthcare. Healthcare organizations must be well prepared to meet these difficulties.

References

Ananda Kumar, S., Mahesh, G., 2021. IoT in Smart Healthcare System.

Bhatt, Y., Bhatt, C., 2017. Internet of things in healthcare. In: Bhatt, C., Dey, N., Ashour, A.S. (Eds.), Internet of Things and Big Data Technologies for Next Generation Healthcare. Springer International Publishing, Cham, pp. 13–33.

Biswas, R., 2022. Outlining big data analytics in health sector with special reference to Covid-19. Wirel. Pers. Commun. 124 (3), 2097–2108. https://doi.org/10.1007/s11277-021-09446-4. Epub 2021 Dec 1. PMID: 34873378; PMCID: PMC8635320.

Chaudhari, S.N., Mene, S.P., Bora, R.M., Somavanshi, K.N., 2020. Role of internet of things (IOT) in pandemic Covid-19 condition. Int. J. Eng. Res. Appl. 10, 57–61. www.ijera.com.

Digiteum Team, 2019. What Connects Internet of Things and Big Data (WWW Document). Digiteum. https://www.digiteum.com/the-internet-of-things-big-data-connected/.

Gupta, R., Bedi, M., Goyal, P., Wadhera, S., Verma, V., 2020. Analysis of COVID-19 tracking tool in India. Digit. Gov. Res. Pract. 1, 1–8.

Haleem, A., Javaid, M., Khan, I.H., Vaishya, R., 2020. Significant applications of big data in COVID-19 pandemic. Indian J. Orthop. 54, 526–528.

Inspirisys Solutions Limited, 2021. 10 IoT Applications in Healthcare Sector You Must Know in 2022 (WWW Document) https://community.nasscom.in/communities/healthtech-and-life-sciences/10-iot-applications-healthcare-sector-you-must-know-2022.

Kajeet Team, 2021. Advantages of IoT in Remote Patient Monitoring (WWW Document) https://www.kajeet.net/advantages-of-iot-in-remote-patient-monitoring/#:~:text=IoT and Remote Patient Monitoring IoT can be, immediately provide medical assistance as and when needed.

Kumar, S., Singh, M., 2019. Big data analytics for healthcare industry: impact, applications, and tools. Big Data Min. Anal. 2, 48–57.

Lirette, C., 2019. What Is the Relationship Between IoT and Big Data? (WWW Document) https://www.soracom.io/blog/what-is-the-relationship-between-iot-and-big-data/.

Malhotra, A., 2018. Internet of Things and Big Data—Better Together (WWW Document) https://www.whizlabs.com/blog/iot-and-big-data/.

Manogaran, G., Lopez, D., Thota, C., Abbas, K.M., Pyne, S., Sundarasekar, R., 2017. Big data analytics in healthcare internet of things. Underst. Complex Syst., 263–284.

Nižetić, S., Šolić, P., López-de-Ipiña González-de-Artaza, D., Patrono, L., 2020. Internet of Things (IoT): opportunities, issues and challenges towards a smart and sustainable future. J. Clean. Prod. 274, 122877.

Raghupathi, W., Raghupathi, V., 2014. Big data analytics in healthcare: promise and potential. Health Inf. Sci. Syst. 2.

Rghioui, A., Lloret, J., Oumnad, A., 2020. Big data classification and internet of things in healthcare. Int. J. E-Health Med. Commun. 11, 20–37.

Singh, A., n.d. IoT—An Introduction [WWW Document]. https://iotpoint.wordpress.com/.

Singh, R.P., Javaid, M., Haleem, A., Suman, R., 2020. Internet of things (IoT) applications to fight against COVID-19 pandemic. Diabetes Metab. Syndr. Clin. Res. Rev. 14, 521–524.

Team, 2020. IoT in Healthcare: Benefits, Use Cases, Challenges, and Future (WWW Document) https://www.intellectsoft.net/blog/iot-in-healthcare/.

TestingXperts Team, 2021. 7 Key Benefits of Big Data Analytics in Healthcare (WWW Document) https://www.testingxperts.com/blog/Big-Data-Analytics-Healthcare.

Undavia, J.N., Patel, A.M., 2020. Big data analytics in healthcare. Int. J. Big Data Anal. Healthc. 5, 19–27.

Wang, Q., Su, M., Zhang, M., Li, R., 2021. Integrating digital technologies and public health to fight Covid-19 pandemic: key technologies, applications, challenges and outlook of digital healthcare. Int. J. Environ. Res. Public Health 18, 6053.

B. Enabling technologies for digital transformation (an example-oriented approach)

4

Blockchain-based electronic health record system in the age of COVID-19

Yang Lu

School of Science, Technology and Health, York St John University, York, United Kingdom

1. Introduction

Digital transformation specifies how the use of technologies can fundamentally transform a particular domain, and the effect created by the application. In the clinical and health sectors, digital transformation such as mobile health devices and wireless technologies has enabled the development of electronic health records (EHRs), wearables, personalized medicine, and telehealth (Faddis, 2018). The COVID-19 pandemic has presented numerous challenges for health providers in frontline care, public health surveillance, and infection prevention and control (Peek et al., 2020). To address these challenges, emerging technologies such as geospatial technology, artificial intelligence, and robotics have been explored as new strategies. Among the 10 key technologies in combating COVID-19 (Kritikos, 2020), blockchain is considered a transparent and efficient means of facilitating effective decision-making, which can lead to faster responses during emergencies (Mbunge et al., 2021).

Blockchain has the potential to become an integral part of the global response to the coronavirus. In addition to monitoring virus spread every day, it can offer the automated platform computational trust for health information exchange among multiple parties. Therefore, using blockchain to streamline case isolation, track drug and medical supplies, manage EHRs, and identify disease symptom patterns can improve the diagnostic accuracy and efficiency (Jadhav and Moosafintavida, 2020). Beyond that, the use of permissioned blockchain systems can help health authorities to tackle the challenge of healthcare interoperability as well as expedite clinical trials by facilitating trustworthy data collection, sharing, reporting, and storage. Compared with traditional surveillance mechanisms in the public health emergency contexts, blockchain's characteristics of encryption as well as decentralized peer-to-peer engagement can help to satisfy the requirements of regulatory compliance (Khubrani and Alam, 2021).

2. Distributed ledger and blockchain

2.1 Distributed ledger technology

A distribution ledger (DL) is "a type of database that is shared, replicated, and synchronized among the members of a network" (Gaynor et al., 2020). The DL can "record transactions such as the exchange of assets or data among participants in the network" (IBM Global Business Services Public Sector Team, 2016). Depending on whether participants need permissions from any entity to make changes to the ledger, DLs can be categorized as the permissioned (private) or permissionless (public) type, as shown in Fig. 1. On a public distribution ledger, each user can have a copy of the ledger and participate in confirming transactions independently, whereas participation in the transaction activities can be restricted on a permissioned distribution ledger.

Having the central party removed can avoid single point of attack in the entire network and thus help to improve the efficiency associated with ledger maintenance and security. Traditional record keeping has always been a centralized process that requires trust in the record keeper. With DL technology, self-interested participants can collectively record verified data in their respective ledgers without relying on a trusted central party. This sets it apart from other technological developments such as cloud computing and data replication, which are commonly used in existing shared ledgers.

2.2 Blockchain

Blockchain was conceptualized in Satoshi Nakamoto's mining computer to solve the double-spending problem by maintaining the order of transactions (Nakamoto, 2008). Initially used to verify Bitcoin transactions, blockchain is structured by a series of transaction blocks and applied to maintain an immutable database within a wide range of domains. Fig. 2 illustrates a simple blockchain with five blocks. Each block contains block data, a hash

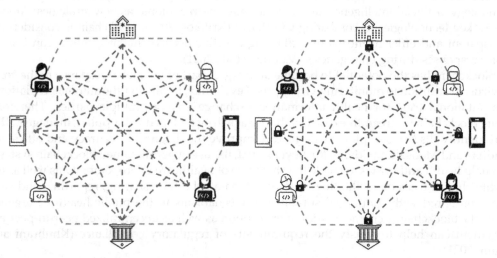

FIG. 1 Distributed ledger technology. (A) Permissionless distributed ledger, (B) permissionless distributed ledger.

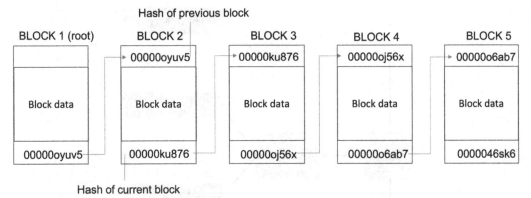

FIG. 2   Blockchain architecture.

of the previous block, and a hash of the data in this block. The dashed lines define the region where the block hash covers. Each block is linked to the previous block and every block on the chain. If any data in a block is changed, the hash from that block and later blocks will become incorrect. The decentralization nature of blockchain enables everyone to have a copy of the chain, and everyone must also make the same change to keep the entire blockchain consistent, which is highly unlikely (i.e., high Byzantine fault tolerance). As new blocks are added, earlier blocks cannot be altered retrospectively by any network member. This is a model of distributed trust, which prevents bad actors from compromising the integrity of the database (Gaynor et al., 2020). Based on the permission given to the network nodes, there are three types of blockchain systems:

- Public (permissionless) blockchain is open to anyone who wants to join any time and acts as a simple node or as a miner for economic rewards. The public blockchain is widely applied in cryptocurrencies such as Bitcoin (Nakamoto, 2008) and Ethereum (Wood, 2014).
- Private (permissioned) blockchain relies on access control, where each participant must obtain an invitation or permission to join. The permissioned blockchain manages privacy by requiring permission from participating entities to confirm transactions (Mahore et al., 2019).
- Consortium (semi-permissioned) blockchain is granted to a group of approved organizations. It is commonly associated with enterprise to improve the transparency and accountability of data-sharing activities (Miyamae et al., 2018).

2.3 Smart contract

The concept of "smart contract" was proposed as computerized transaction protocols that execute the terms of a contract. A cryptocurrency is essentially "a decentralized system for interacting with virtual money in a shared global ledger" (Giancaspro, 2017). Implemented as an instance of a computer program running on blockchain, smart contracts are constructed upon an underlying cryptocurrency platform. Conceptually, a contract can be treated as a special trusted third party (TTP) and the contract's code is executed whenever it receives a message, either from a user or from another contract (Delmolino et al., 2016). A smart contract can serve as motivation for the useful aspects of smart contracts as financial instruments

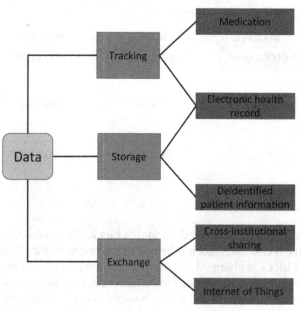

FIG. 3 Information exchange among health record systems.

(Zou et al., 2019). Zhang et al. (2018a) applied the smart contract-enabled blockchain technology to implement distributed and trustworthy access control for the Internet of Things (IoT). Combined with the symmetric encryption algorithm (SEA), a blockchain-based access control scheme was proposed for privacy-preserving distributed IoT data stored in the management servers (Shi et al., 2021).

3. Digital health and blockchain

Managing a large amount of data collected from health systems can be challenging (Kruse et al., 2016). Blockchain is becoming recognized in the health sectors for securing records by validating the transactions created. Fig. 3 depicts some applications that can allow the healthcare industry to improve data exchange across all different areas, including record tracking, exchange, and storage (Gaynor et al., 2020). For instance, with the data as input, patients and organizations can apply clinical and health services (Ribitzky et al., 2018). Blockchain-based methods can be applied as a solution to deal with security breaches in the entire life cycle of medical record management. The private and public key systems built in blockchain allow for trust to be established to transfer data in a manner that is traceable and secure (Kuo et al., 2017; Park et al., 2019). Thanks to the traceability and transparency features, blockchain architecture can create a more efficient and effective way to manage high-value items, medical devices, and durable medical equipments along the supply chain. Besides, blockchain can be deployed to keep participants informed and engaged in the evolution of the records in the health environment. As shown in Fig. 4, smart contract can serve the operation workflows, health tracking, and data management in the medical contexts.

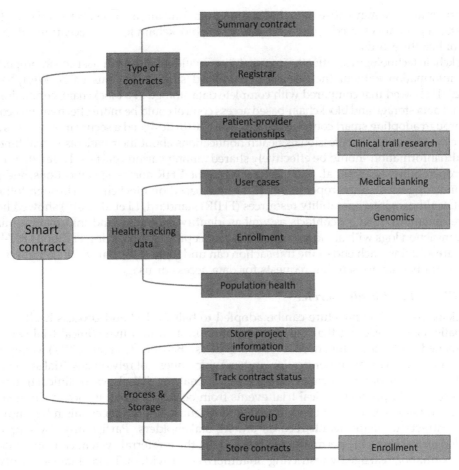

FIG. 4 Smart contract used in a digital health system.

In this section, we analyze blockchain applications in health sectors with the major focus on data management, clinical and public health services, as well as medical supply chains. More importantly, we discuss the added values that blockchain can bring to digital health systems in terms of security and data privacy.

3.1 Health data and services

3.1.1 Data management

The pathway to digital health is a transformation of traditional practices of health care, including the access to electronic health records, remote health monitoring, user-centric patient portals, wearables, data interoperation for analytics, and so on (Sust et al., 2020). Sanmarchi et al. (2021) shed light on cloud-based and DLT solutions, emphasizing the importance of digital solutions in giving patients full control over their health records. This will enhance organizational processes and increase the investment return. Considering that institution-based record maintenance may cause data thefts, manipulation, and misuses, Harshini et al. (2019) suggested a patient-centered approach by adopting blockchain for the decentralized maintenance of health records.

Blockchain technology can offer the immutability, confidentiality, and user access properties of stored information without the needs of centralized storage. On this basis, Westphal and Seitz (2021) showed that compared with complete data storage in a blockchain, combining conventional data storage and blockchain-based access control could be more effective and economical. Through adopting smart contracts, Griggs et al. (2018) designed a secure monitoring system which can send medical professionals health notifications about their patients in real-time.

Health information should be effectively shared among various parties. To satisfy the need of interoperability, Reegu et al. (2021a,b) analyzed the EHR norms, specifications, and problems, and suggested an interoperable EHR system based on blockchain. Through following the fast healthcare interoperability resources (FHIR) standard, Li et al. (2019) showed how to capture semantics of smart contacts as well as identify anomalies and information misuses from transaction logs with standard ontological concepts. By supporting FHIR on DLT-based healthcare systems, each end of the transaction can understand the data while identifying the behaviors of bad actors sending requests for data access or usage.

3.1.2 Clinical and health service

Blockchain-based architecture can be adopted to help collect and process health records from patients. Considering that the authentication system in a live clinical trial should be strengthened to remove the need of third parties, Benchoufi et al. (2017) suggested a blockchain-based patient authentication system which does not rely on any trial stakeholder. Benefiting from the decentralization and traceability nature, data flows in clinical trials could be tracked to help prevent clinical trial events from occurring in an incorrect chronological order. Chenthara et al. (2020) introduced a patient-centric referral mechanism by employing smart contracts to share health records among stakeholders. Particularly, by employing Hyperledger Fabric as the permissioned blockchain, the e-referral system can eliminate sensitive information leakage by restricting unauthorized providers. To facilitate authority verification, abnormal alerting, attack resistance, and traceability query, Xu et al. (2021) proposed a distributed, structural health monitoring (SHM) system where node downtime cannot cause data loss or affect the operation of the system. To enhance the ability of insurance companies to obtain information, Chen et al. (2021) adopted blockchain technology for online medical insurance claims. With smart contract and data sharing, it supports automatic claims settlements, customer identity verification, and money laundering prevention.

3.1.3 Medical supply chain

In the healthcare environment, blockchain has the power to support cost tracking along supply chains. Fig. 5 illustrates opportunities for blockchain in the medical supply chain (Gaynor et al., 2020). This enables real-time tracking of high-value items and rapid detection of any errors in order to avoid errors and improve the use of limited resources of the supply chain (Kumar and Tripathi, 2019; Mann et al., 2018). One example is the implementation of blockchain technologies for organ transportation. By tracking the organ on its journey from the donor to recipient, it can improve efficiency and allow for innovation in this process such as the use of drones for organ delivery (Chavez et al., 2020; Goodman, 2014). This will be further discussed in Section 4.

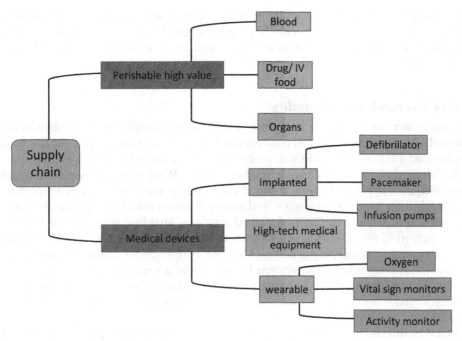

FIG. 5 Blockchain applications in medical supply chains.

3.2 Privacy and security

EHR systems allow storing, sharing, managing, controlling, and maintaining patient information among healthcare providers. Due to the existing issues such as privacy leakage and unexpected data access, blockchain and smart contracts can be adopted in EHR management to ensure the confidentiality and utility while protecting sensitive patient data (Alsamhi et al., 2021).

3.2.1 Trust management

Authentication and data integrity can help strengthen user trust in digital health applications. Brogan et al. (2018) demonstrated a masked messaging module with the IOTA (IOTA, 2022) protocols applied to share, store, and retrieve encrypted activity data through a tamper-proof distributed ledger. Chelladurai et al. (2021) designed patient-centered smart contracts for patient registration, health record creation, access, update, storage, and sharing for home care and future care services. To ensure patients and providers can manage the personal health information (PHI) data maintained by different healthcare providers, Rahmadika and Rhee (2018) proposed a conceptual model based on blockchain technology. Due to the immutable characteristic, it allows patients to get a single view of their PHI data while guarantee data integrity. Based on the traceability property, Gong et al. (2019) proposed personal health-related application data provenance to ensure the right confirmation. In addition, it can be used to identify the rights to use personal health data, which can further ensure the data security. The dissemination of patients' medical records may lead to privacy leakage. To address this while sharing medical data among data custodians in a trustless environment, Xia et al. (2017) proposed MeDShare to record data transitions and behaviors in a tamper-proof manner. As a result, it can effectively track and revoke access when permission violation is detected.

3.2.2 Access control: Confidentiality

Blockchain technology can be applied to build patient-centric access control systems to provide individuals secure and full data access. For instance, Ekblaw et al. (2016) proposed a decentralized EHR system with comprehensive, immutable log and easy access to their medical information across various medical sites. With the privileges to aggregate (anonymized) data as reward, medical stakeholders are incentivized to participate in the network as blockchain "miners" to sustain and secure the network via the proof of work. Later, Cunningham and Ainsworth (2018) designed a system enabling fine-grained personalized control of patient EHRs, allowing individuals to specify when and by whom their records can be accessed for research purposes. The use of the smart contract-based Ethereum blockchain technology allows the system to operate in a verifiably secure, trustless, open and auditable environment. Fig. 6 shows a blockchain-based health system that enables secure medical data storage and exchange based on granular access controls (Panigrahi et al., 2022).

As digital health shifts from the volume-based care to value-based care, cross-institutional medical data exchange becomes significant in the healthcare domain. Duong-Trung et al. (2020) proposed a patient-centered care mechanism by using smart contracts and blockchain technology. Patients can easily manage their data while tracing any change made to their data by other care providers. By integrating the FHIR with smart contracts for patient-generated data sharing, Hylock and Zeng (2019) designed HealthChain to allow patients and providers to access consistent and comprehensive medical records. Similarly, Zhang et al. (2018b) developed FHIRChain to share clinical data securely. As shown in Fig. 7, FHIRChain is built upon a token-based permission model by using the encrypt-then-authenticate method (Sookhak et al., 2021). Particularly, EHRs will be signed with the data owner's private key and then encrypted with the requester's digital health identities (public key) for ubiquitous access.

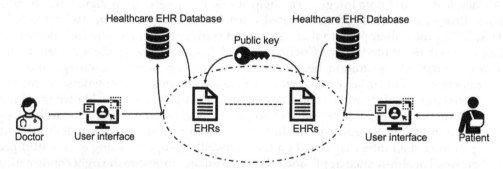

FIG. 6 Block diagram for the proposed system.

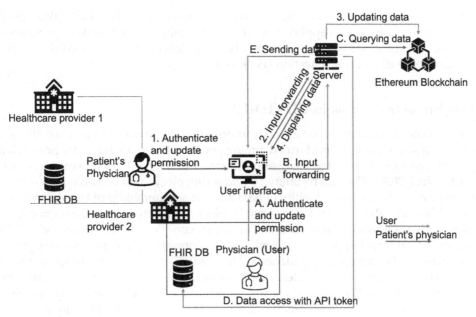

FIG. 7   Access control in FHIRChain.

The pervasive IoT devices allow new opportunities for healthcare such as remote patient monitoring. However, there are security and privacy considerations in data transmission, from these devices to the back-end server, and across heterogeneous IoT networks. Saweros and Song (2019) proposed the utilization of distributed transaction ledger (DTL) technology to facilitate communication among patients and care providers. This approach can resolve the data fragmentation issue and provide a relatively complete picture of each patient, supporting personalized exercise tracking and monitoring of biometric data. The adoption of Tangle can ensure sensitive data protection. Ray et al. (2021) proposed a privacy-preserving scheme named BIoTHR based on blockchain and swarm exchange techniques to facilitate seamless and secure EHR information transmission through the secure, peer-to-peer communication among swarm nodes.

4. Blockchain technologies in combating COVID-19

The inherent attributes of blockchain technology make it suitable in deploying medical applications to combat the pandemic situation efficiently. As introduced, the major contribution of blockchain is to preserve the privacy in EHR information while ensuring the immutability, transparency, and traceability in data exchange. While fighting COVID-19, healthcare professionals and individuals can utilize these blockchain applications to exchange personal data in EHRs while ensuring security and privacy (Sharma et al., 2020). Besides, the decentralized service architecture can address most of the limitations associated with centralized computational systems (Kalla et al., 2020). Toward the strategic needs in the COVID-19 crisis, this section reviews blockchain applications of various healthcare and medical purposes such as contact tracing, vaccine proof, supply chain, health monitoring, patient record privacy, test result sharing as well as early detection (Peek et al., 2020).

4.1 Combating trust crisis in COVID-19

In the absence of reliable data or accurate information, there needs to be a "trustless" system where transactions can be performed among people who do not have any prior relationship yet still can validate the objectivity and principles of the medium in which transactions occur (Khurshid, 2020). This can be offered by adopting blockchain technology to ensure the transparency of contracts, immutability of data and the accountability of transactions among strangers (Xia et al., 2017). For instance, a public blockchain network can guarantee full controls of individual's information while maintaining the complete privacy (Kuo et al., 2017). By using a blockchain-based approach, patient health information such as blood oxygen levels and heart rates can be managed by patients and hospitals without requesting access from centralized third-party databases. Thus, test results would be more accurate and the transmission would be more efficient. Particularly, confirmed positive cases can be screened by collecting and sharing personal information among health collaborators while ensuring the personal data protection against robust access control policies (Abid et al., 2020). Ransomware attacks can cause negative impacts to hospital operations such as delaying the access to health services and significant economic loss. As hospitals were targeted and attacked during the pandemic, ransomware issues were reported by hospitals and testing facilities in Germany, the United States, and the United Kingdom (Eddy and Perlroth, 2020). To address these challenges, artificial intelligence and blockchain technologies were employed to prevent cyberattacks effectively (Firdaus et al., 2018). At the beginning of the pandemic, the World Health Organization (WHO) reported that increasing fake news and deepfakes related to the COVID-19 pandemic worsened the crisis of trust in government institutions and public health agencies, causing serious threats to public health. Jing and Murugesan (2020) proposed a trust index model using blockchain technology to trace the information sources and evaluate based on the author's credibility. Blockchain features such as transparency, immutability, and decentralization make it suitable for social media ecosystems to protect user data privacy.

4.2 Health monitoring

4.2.1 Vaccination and immunity certification

As COVID-19 infection can be asymptomatic, it can be difficult to prevent virus transmission. Despite various containment policies such as city lockdowns, mandatory masking, and contact tracing, COVID-19 vaccine presents a new option to slow down the pandemic outbreak. Posttrial monitoring is of utmost importance for the safety and health of vaccinated people. Tsoi et al. (2021) introduced a vaccine passport concept implemented with blockchain technology that can be a promising tool for health monitoring while protecting personal privacy. Bansal et al. (2020) utilized blockchain to mitigate the falsification of test reports, which allowed people to safely meet others who have the immunity-based license. The blockchain-based system offers an opportunity to prioritize individuals for antibody testing via contact tracing. Timely and accurate testing can be paramount to effective pandemic responses. Hasan et al. (2020) proposed digital medical passports and immunity certificates for COVID-19 test takers. Based on the Ethereum blockchain written and tested successfully to maintain a medical identity for test takers, smart contracts can help prompt trusted responses by medical authorities.

With the inherent value of blockchain, a decentralized vaccination passport can be developed as an incentive system with a customized cryptocurrency to reward users (de los Santos Nodar and Fernández Caramés, 2021). The lack of legal identity made some poor migrants and refugees vulnerable during the pandemic (Shuaib et al., 2021). Self-sovereign identity (SSI) is a form of distributed digital identity that can provide an immutable identity with full user control and interoperability features (Sharma, 2021; Ledger Insights, 2020). A blockchain-based SSI model cannot only provide migrants and refugees an effective legal identity and include them in government welfare schemes, it also can be used to identify people who have been vaccinated or tested. These findings would encourage people to return to workplace and resume their activities confidently (Teilen, 2020).

4.2.2 Contact tracing

Unprecedented social pressure is pushing toward the adoption of contact tracing systems in response to the COVID-19 pandemic. Due to its inherent features, blockchain can be leveraged to track the spread of the coronavirus infection globally. In addition, it can facilitate the fast-tracking of clinical trials, health records, and support any fundraising activities or donations (Ricci et al., 2021). Due to blockchain's tamper-proof nature, complete privacy can be guaranteed among users on the blockchain. Based upon the blockchain architecture, the Pronto-C2 system can provide stronger protection from complete transparency and resilience through full decentralization (Avitabile et al., 2020). Through the leverage of blockchain, smart contracts, and Bluetooth, Song et al. (2021) designed a global COVID-19 information sharing and risk notification system to share consistent and non-tampered contact tracing information while protecting user privacy. Xu et al. (2020) presented a blockchain-enabled privacy-preserving contact tracing scheme, BeepTrace. Through adopting blockchain to bridge the user/patient and authorized solvers, it can desensitize user ID from location information so as to ensure higher security and privacy. This solution provides a timely framework for governments, authorities, companies, and research institutes worldwide to develop a trusted platform for tracing information sharing to win the fight against the COVID-19 pandemic. Lv et al. (2020) proposed a decentralized and permissionless blockchain protocol, Bychain, to protect data security and location privacy while deploying short-range communication (SRC) IoT witnesses to monitor large areas. Regardless of decoupling personal identities from the ownership of on-chain location, it can allow the location data owner to claim their ownership without revealing the private key to others.

Analyzing data collected from multiple governmental and hospital networks can help to identify high-risk individuals and positive cases in the process of contact tracing (Garg et al., 2020). Fernández-Caramés et al. (2020) presented an IoT occupancy system to estimate the occupancy level of public spaces through deploying IoT devices in the monitored areas. With a decentralized traceability subsystem built on a blockchain, it can guarantee the availability and immutability of collected information.

4.2.3 Solutions to social distancing

Governments fought the COVID-19 pandemic using unmanned aerial vehicles (UAVs) to monitor social distancing and sanitize affected areas. For instance, UAVs can measure COVID-19 symptoms such as heart rate, blood pressure, and body temperature. Based on that, COVID-19 patients can be tracked and quarantined for a specific period. Gupta et al. (2020) designed a multi-swarming application of UAVs to monitor health and social distancing, send notifications to the ground stations and inform the sanitization UAV about the affected area for the purpose of disinfection. The application layer stores the UAV control information, that is, UAV movement through remote control in the blockchain network. Benefits of developing a blockchain-based multiswarming UAV network include data immutability, traceability, transparency, reliability, security, and privacy.

Remotely monitored patients equipped with wearable devices can generate and store data to decentralized applications, as shown in Fig. 8. Collected sensor data can be evaluated to determine whether an alert should be issued to master devices. After verification, the alert will be written onto the blockchain, and detailed information will be stored in the EHRs connected to the blockchain (Zheng et al., 2020).

Robotics become a promising technology for multitasking (Allam and Jones, 2020). Through reducing human interaction, monitoring and delivering goods to combat COVID-19, robots are required to provide healthcare services in the quarantine area (Yang et al., 2020). Alsamhi and Lee (2020) proposed a blockchain-based solution for multirobot controlling, which can avoid collisions during operation in uncertain conditions and environment information exchanges. The combined application of blockchain and multirobot controlling can improve task formulation, decision making, authentication, automation, and action validation.

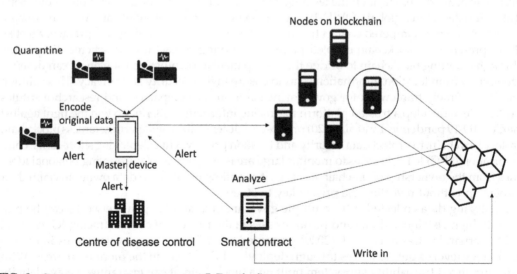

FIG. 8 Smart monitoring system structure (IoT method).

4.3 Early diagnosis and intervention

4.3.1 Efficient information exchange

The COVID-19 pandemic has added extra pressure to existing health information exchange (HIE) systems, forcing the medical providers worldwide to share EHRs effectively and securely among remote care locations (Zheng et al., 2019). To improve the data fluidity for remotely sharing medical data beyond local data storage, Christodoulou et al. (2020) have presented a blockchain-based HIE framework where audit trails of information flows are recorded with the use of a smart contract on an Ethereum blockchain. Instead of requesting access to central storage services, the framework allows immutable data exchange and thus overcomes the interoperability barrier of different hospital systems. Working with tech companies, the WHO has introduced MiPasa (MiPasa, 2022), a platform built on Hyperledger Fabric to cross-reference location data with health information. In addition to ensuring patient privacy, it can help monitoring local and global infections during the pandemic.

4.3.2 Outbreak prediction and prevention

Artificial intelligence and deep learning methods are applied for COVID-19 symptom prediction. With blockchain technology maintaining patient records, predictive analytics and big data technology can assess the situation and trace the spread of the virus (Ahir et al., 2020). For instance, blockchain has been applied increasingly to create a proper basis for an evidence-based decisional process of health management. The use of blockchain in combination with artificial intelligence allows the creation of generalizable predictive systems that can help to control the pandemic risks on national territory (Fusco et al., 2020). Kumar et al. (2021) have proposed a framework to collect heterogenous data from different hospitals and train a global deep learning model by using blockchain-based federated learning. In particular, blockchain technology can authenticate the data, and blockchain-based federated learning can preserve the privacy of the organization. The proposed framework can utilize up-to-date data, which improves the recognition of CT images and detection of COVID-19 patients.

4.4 Medical supply chain

Drones are effective tools used in the medical delivery at a low cost and less human involvement. Singh et al. (2020) proposed a communication model within the internet of drones (IoD) environment where participating entities use a permissioned private blockchain to ensure the participation of only trusted entities to form a peer-to-peer (P2P) network and perform the actions authorized to them. Fig. 9 shows a blockchain-based solution for distributed decision-making required in multidrone collaboration to fight COVID-19. The information of action is stored in the block and accessible by public drones in the group. Each drone requires the agreement of other drones to act accordingly in every subtask performed (Alsamhi et al., 2021).

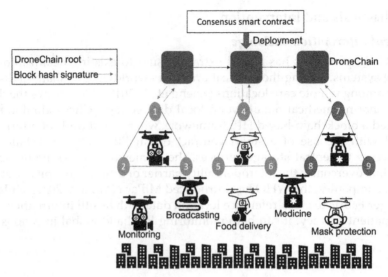

FIG. 9 Multidrone collaboration in combating Covid-19.

5. Summary

In the clinical and health sectors, digital transformation has enabled the development of EHRs, wearables, personalized medicine, and telehealth. Particularly, blockchain technology can play an integral role in the fight against COVID-19. Due to its inherent properties of immutability, confidentiality, trust, traceability, and decentralization, blockchain technology can be used in a new workflow development based on improved protocols in security and privacy protection.

Regardless of the significant progress in fighting COVID-19, it is necessary to explore and implement blockchain applications for health monitoring, diagnosing, screening, tracking, surveillance, and raising awareness. Future studies can be planned to satisfy the need of accurate prediction of early diagnosis of COVID-19. Also, the future work should focus on the ethical framework and adoption of emerging technologies. For instance, while using COVID-19 contact tracing applications to monitor the infection rate, it is critical to ensure the security and privacy of people's trajectories when sharing them. Blockchain has been continuously recognized in securing record sharing between two parties, improving data integrity and trustworthiness by validating whether the transactions happened. According to Mbunge et al. (2021), there are challenges in the implementation of blockchain-based health systems:

- Lack of technical solutions to integrate existing blockchain with health information systems.
- Lack of awareness about applying blockchain in health systems.
- Limited scalability in the application programming interfaces (APIs) for bridging blockchain with EHR systems.
- Ethical issues for integrating blockchain into health systems.

- Unclear regulations and standards on integration.
- Data leakage through blockchain API and cloud-based platforms.

Although blockchain-based health systems are developed to combat COVID-19, further studies are needed to assess the performance of on-chain data management (Abd-Alrazaq et al., 2021). With the hash and metadata stored on blockchain, it can support data integrity and access control through off-chain communication channels (Staples et al., 2017). While adopting blockchain in combination with robotics, it is challenging to evaluate robot sensing capabilities and data quality, which can cause inaccuracy of IoT data collection. Blockchain-based medical delivery systems can be developed to validate data stamps from multirobot collaborations, and thus ensure essential pharmaceutical service during the pandemic.

6. Assignment

Choose a topic in the following list and write a 1500- to 2000-word research essay demonstrating how blockchain technology can be used for patients' privacy protection in the clinical and medical contexts. The report should cite 10-12 scientific articles.

- Informed consent service.
- Health screening service.
- Health insurance service.
- Smart nursing home.
- Medical supply chain.

References

Abd-Alrazaq, A.A., Alajlani, M., Alhuwail, D., Erbad, A., Giannicchi, A., Shah, Z., Househ, M., 2021. Blockchain technologies to mitigate COVID-19 challenges: a scoping review. Comput. Methods Programs Biomed. Update 1, 100001.

Abid, A., Cheikhrouhou, S., Kallel, S., Jmaiel, M., 2020. How blockchain helps to combat trust crisis in COVID-19 pandemic? Poster abstract. In: Proceedings of the 18th Conference on Embedded Networked Sensor Systems, pp. 764–765.

Ahir, S., Telavane, D., Thomas, R., 2020. The impact of Artificial Intelligence, Blockchain, Big Data and evolving technologies in Coronavirus Disease-2019 (COVID-19) curtailment. In: 2020 International Conference on Smart Electronics and Communication (ICOSEC). IEEE, pp. 113–120.

Allam, Z., Jones, D.S., 2020. On the coronavirus (COVID-19) outbreak and the smart city network: universal data sharing standards coupled with artificial intelligence (AI) to benefit urban health monitoring and management. In: Healthcare. vol. 8. Multidisciplinary Digital Publishing Institute, p. 46. No. 1.

Alsamhi, S.H., Lee, B., 2020. Blockchain for Multi-Robot Collaboration to Combat COVID-19 and Future Pandemics. arXiv preprint arXiv:2010.02137.

Alsamhi, S.H., Lee, B., Guizani, M., Kumar, N., Qiao, Y., Liu, X., 2021. Blockchain for decentralized multi-drone to combat COVID-19 and future pandemics: framework and proposed solutions. Trans. Emerg. Telecommun. Technol., e4255.

Avitabile, G., Botta, V., Iovino, V., Visconti, I., 2020. Towards defeating mass surveillance and SARS-CoV-2: the Pronto-C2 fully decentralized automatic contact tracing system. IACR Cryptol. ePrint Arch. 2020, 493.

Bansal, A., Garg, C., Padappayil, R.P., 2020. Optimizing the implementation of COVID-19 "immunity certificates" using blockchain. J. Med. Syst. 44 (9), 1–2.

Benchoufi, M., Raphael, P., Philippe, R., 2017. Blockchain protocols in clinical trials: transparency and traceability of consent. F1000Research 6.

Brogan, J., Baskaran, I., Ramachandran, N., 2018. Authenticating health activity data using distributed ledger technologies. Comput. Struct. Biotechnol. J. 16, 257–266.

Chavez, N., Kendzierskyj, S., Jahankhani, H., Hosseinian, A., 2020. Securing transparency and governance of organ supply chain through blockchain. In: Policing in the Era of AI and Smart Societies. Springer, Cham, pp. 97–118.

Chelladurai, M.U., Pandian, S., Ramasamy, K., 2021. A blockchain based patient centric EHR storage and integrity management for e-health systems. Health Policy Technol., 100513.

Chen, C.L., Deng, Y.Y., Tsaur, W.J., Li, C.T., Lee, C.C., Wu, C.M., 2021. A traceable online insurance claims system based on blockchain and smart contract technology. Sustainability 13 (16), 9386.

Chenthara, S., Ahmed, K., Wang, H., Whittaker, F., 2020. A novel blockchain based smart contract system for eReferral in healthcare: HealthChain. In: International Conference on Health Information Science. Springer, Cham, pp. 91–102.

Christodoulou, K., Christodoulou, P., Zinonos, Z., Carayannis, E.G., Chatzichristofis, S.A., 2020. Health information exchange with blockchain amid COVID-19-like pandemics. In: 2020 16th International Conference on Distributed Computing in Sensor Systems (DCOSS). IEEE, pp. 412–417.

Cunningham, J., Ainsworth, J., 2018. Enabling patient control of personal electronic health records through distributed ledger technology. Stud. Health Technol. Inform. 245, 45–48.

Delmolino, K., Arnett, M., Kosba, A., Miller, A., Shi, E., 2016. Step by step towards creating a safe smart contract: lessons and insights from a cryptocurrency lab. In: International Conference on Financial Cryptography and Data Security. Springer, Berlin, Heidelberg, pp. 79–94.

Duong-Trung, N., Son, H.X., Le, H.T., Phan, T.T., 2020. Smart care: integrating blockchain technology into the design of patient-centered healthcare systems. In: Proceedings of the 2020 4th International Conference on Cryptography, Security and Privacy, pp. 105–109.

Eddy, M., Perlroth, N., 2020. Cyber Attack Suspected in German Woman's Death. The New York Times. https://www.nytimes.com/2020/09/18/world/europe/cyber-attack-germany-ransomeware-death.html. (Accessed 1 October 2020).

Ekblaw, A., Azaria, A., Halamka, J.D., Lippman, A., 2016. A case study for blockchain in healthcare: "MedRec" prototype for electronic health records and medical research data. In: Proceedings of IEEE Open and Big Data Conference. vol. 13, p. 13.

Faddis, A., 2018. The digital transformation of healthcare technology management. Biomed. Instrum. Technol. 52 (s2), 34–38.

Fernández-Caramés, T.M., Froiz-Míguez, I., Fraga-Lamas, P., 2020. An IoT and blockchain based system for monitoring and tracking real-time occupancy for covid-19 public safety. In: Engineering Proceedings. vol. 2. Multidisciplinary Digital Publishing Institute, p. 67. No. 1.

Firdaus, A., Anuar, N.B., Razak, M.F.A., Hashem, I.A.T., Bachok, S., Sangaiah, A.K., 2018. Root exploit detection and features optimization: mobile device and blockchain based medical data management. J. Med. Syst. 42 (6), 1–23.

Fusco, A., Dicuonzo, G., Dell'Atti, V., Tatullo, M., 2020. Blockchain in healthcare: insights on COVID-19. Int. J. Environ. Res. Public Health 17 (19), 7167.

Garg, C., Bansal, A., Padappayil, R.P., 2020. COVID-19: prolonged social distancing implementation strategy using blockchain-based movement passes. J. Med. Syst. 44 (9), 1–3.

Gaynor, M., Tuttle-Newhall, J., Parker, J., Patel, A., Tang, C., 2020. Adoption of blockchain in health care. J. Med. Internet Res. 22 (9), e17423.

Giancaspro, M., 2017. Is a 'smart contract' really a smart idea? Insights from a legal perspective. Computer Law Security Rev. 33 (6), 825–835.

Gong, J., Lin, S., Li, J., 2019. Research on personal health data provenance and right confirmation with smart contract. In: 2019 IEEE 4th Advanced Information Technology, Electronic and Automation Control Conference (IAEAC). vol. 1. IEEE, pp. 1211–1216.

Goodman, L.M., 2014. Tezos: A Self-Amending Crypto-Ledger Position Paper. vol. 3, pp. 1–18.

Griggs, K.N., Ossipova, O., Kohlios, C.P., Baccarini, A.N., Howson, E.A., Hayajneh, T., 2018. Healthcare blockchain system using smart contracts for secure automated remote patient monitoring. J. Med. Syst. 42 (7), 1–7.

Gupta, R., Kumari, A., Tanwar, S., Kumar, N., 2020. Blockchain-envisioned softwarized multi-swarming uavs to tackle covid-i9 situations. IEEE Netw. 35 (2), 160–167.

Harshini, V.M., Danai, S., Usha, H.R., Kounte, M.R., 2019. Health record management through blockchain technology. In: 2019 3rd International Conference on Trends in Electronics and Informatics (ICOEI). IEEE, pp. 1411–1415.

Hasan, H.R., Salah, K., Jayaraman, R., Arshad, J., Yaqoob, I., Omar, M., Ellahham, S., 2020. Blockchain-based solution for COVID-19 digital medical passports and immunity certificates. IEEE Access 8, 222093–222108.

Hylock, R.H., Zeng, X., 2019. A blockchain framework for patient-centered health records and exchange (HealthChain): evaluation and proof-of-concept study. J. Med. Internet Res. 21 (8), e13592.

IBM Global Business Services Public Sector Team, 2016. Blockchain: The Chain of Trust and its Potential to Transform Healthcare—Our Point of View. Office of the National Coordinator for Health Information.

IOTA, 2022. https://www.iota.org/. Accessed 20 Feb.

Jadhav, V.D., Moosafintavida, D.S., 2020. Blockchain in healthcare industry and its application and impact on Covid 19 digital technology transformation. Mukt Shabd J. 9, 479.

Jing, T.W., Murugesan, R.K., 2020. Protecting data privacy and prevent fake news and deepfakes in social media via blockchain technology. In: International Conference on Advances in Cyber Security. Springer, Singapore, pp. 674–684.

Kalla, A., Hewa, T., Mishra, R.A., Ylianttila, M., Liyanage, M., 2020. The role of blockchain to fight against COVID-19. IEEE Eng. Manag. Rev. 48 (3), 85–96.

Khubrani, M.M., Alam, S., 2021. A detailed review of blockchain-based applications for protection against pandemic like COVID-19. Telkomnika 19 (4), 1185–1196.

Khurshid, A., 2020. Applying blockchain technology to address the crisis of trust during the COVID-19 pandemic. JMIR Med. Inform. 8 (9), e20477.

Kritikos, M., 2020. Ten Technologies to Fight Coronavirus, European Parliamentary Research Service. European Parliamentary Research Service (EPRS). Accessed 4 Feb 2022.

Kruse, C.S., Goswamy, R., Raval, Y., Marawi, S., 2016. Challenges and opportunities of big data in health care: a systematic review. JMIR Med. Inform. 4 (4), e38. https://doi.org/10.2196/medinform.5359.

Kumar, R., Khan, A.A., Kumar, J., Zakria, A., Golilarz, N.A., Zhang, S., Wang, W., 2021. Blockchain-federated-learning and deep learning models for covid-19 detection using CT imaging. IEEE Sensors J. 21.

Kumar, R., Tripathi, R., 2019. Traceability of counterfeit medicine supply chain through Blockchain. In: 2019 11th International Conference on Communication Systems & Networks (COMSNETS). IEEE, pp. 568–570.

Kuo, T.T., Kim, H.E., Ohno-Machado, L., 2017. Blockchain distributed ledger technologies for biomedical and health care applications. J. Am. Med. Inform. Assoc. 24 (6), 1211–1220.

Ledger Insights, 2020. Immunity Certificates Don't Have to Give Up Our Data Privacy. Here's why. Ledger Insights—enterprise blockchain, Ledger Insights. https://www.ledgerinsights.com/immunity-certificates-dont-have-to-give-up-our-data-privacy/. Accessed Jan. 22, 2022.

Li, M., Xia, L., Seneviratne, O., 2019. Leveraging standards based ontological concepts in distributed ledgers: a healthcare smart contract example. In: 2019 IEEE International Conference on Decentralized Applications and Infrastructures (DAPPCON). IEEE, pp. 152–157.

de los Santos Nodar, M.A., Fernández Caramés, T.M., 2021. COVID-19 digital vaccination passport based on blockchain with its own cryptocurrency as a reward and mobile app for its use. Engi. Proc. 7 (1), 35.

Lv, W., Wu, S., Jiang, C., Cui, Y., Qiu, X., Zhang, Y., 2020. Towards large-scale and privacy-preserving contact tracing in COVID-19 pandemic: a blockchain perspective. IEEE Trans. Network Sci. Eng. 9.

Mahore, V., Aggarwal, P., Andola, N., Venkatesan, S., 2019. Secure and privacy focused electronic health record management system using permissioned blockchain. In: 2019 IEEE Conference on Information and Communication Technology. IEEE, pp. 1–6.

Mann, S., Potdar, V., Gajavilli, R.S., Chandan, A., 2018. Blockchain technology for supply chain traceability, transparency and data provenance. In: Proceedings of the 2018 International Conference on Blockchain Technology and Application, pp. 22–26.

Mbunge, E., Akinnuwesi, B., Fashoto, S.G., Metfula, A.S., Mashwama, P., 2021. A critical review of emerging technologies for tackling COVID-19 pandemic. Human Behav. Emerg. Technol. 3 (1), 25–39.

MiPasa, 2022. https://mipasa.org. Accessed 20 Feb.

Miyamae, T., Honda, T., Tamura, M., Kawaba, M., 2018. Performance improvement of the consortium blockchain for financial business applications. J. Digital Bank. 2 (4), 369–378.

Nakamoto, S., 2008. Bitcoin: a peer-to-peer electronic cash system. Decentral. Business Rev., 21260.

Panigrahi, A., Nayak, A.K., Paul, R., 2022. HealthCare EHR: a blockchain-based decentralized application. Int. J. Inform. Syst. Supply Chain Manag. (IJISSCM) 15 (3), 1–15.

B. Enabling technologies for digital transformation (an example-oriented approach)

Park, Y.R., Lee, E., Na, W., Park, S., Lee, Y., Lee, J.H., 2019. Is blockchain technology suitable for managing personal health records? Mixed-methods study to test feasibility. J. Med. Internet Res. 21 (2), e12533.

Peek, N., Sujan, M., Scott, P., 2020. Digital health and care in pandemic times: impact of COVID-19. BMJ Health Care Inform. 27 (1).

Rahmadika, S., Rhee, K.H., 2018. Blockchain technology for providing an architecture model of decentralized personal health information. Int. J. Eng. Business Manag. 10. 1847979018790589.

Ray, P.P., Chowhan, B., Kumar, N., Almogren, A., 2021. BIoTHR: electronic health record servicing scheme in IoT-blockchain ecosystem. IEEE Internet Things J. https://doi.org/10.1109/jiot.2021.3050703.

Reegu, F.A., Al-Khateeb, M.O., Zogaan, W.A., Al-Mousa, M.R., Alam, S., Al-Shourbaji, I., 2021a. Blockchain-based framework for interoperable electronic health record. Ann. Roman. Soc. Cell Biol., 6486–6495.

Reegu, F.A., Mohd, S., Hakami, Z., Reegu, K.K., Alam, S., 2021b. Towards trustworthiness of electronic health record system using blockchain. Ann. Roman. Soc. Cell Biol. 25 (6), 2425–2434.

Ribitzky, R., Clair, J.S., Houlding, D.I., McFarlane, C.T., Ahier, B., Gould, M., Clauson, K.A., 2018. Pragmatic, inter-disciplinary perspectives on blockchain and distributed ledger technology: paving the future for healthcare. Blockchain Healthc. Today 1.

Ricci, L., Maesa, D.D.F., Favenza, A., Ferro, E., 2021. Blockchains for covid-19 contact tracing and vaccine support: a systematic review. IEEE Access 9, 37936–37950.

Sanmarchi, F., Toscano, F., Fattorini, M., Bucci, A., Golinelli, D., 2021. Distributed solutions for a reliable data-driven transformation of healthcare management and research. Front. Public Health 9, 944.

Saweros, E., Song, Y.T., 2019. Connecting personal health records together with EHR using tangle. In: 2019 20th IEEE/ACIS International Conference on Software Engineering, Artificial Intelligence, Networking and Parallel/Distributed Computing (SNPD). IEEE, pp. 547–554.

Sharma, T., 2021. Widespread Adoption of Self-Sovereign Identity in the Wake of COVID-19. Blockchain Council. https://www.blockchain-council.org/blockchain/widespread-adoption-of-self-sovereign-identity-in-the-wake-of-covid-19/. (Accessed 14 January 2021).

Sharma, A., Bahl, S., Bagha, A.K., Javaid, M., Shukla, D.K., Haleem, A., 2020. Blockchain technology and its applications to combat COVID-19 pandemic. Res. Biomed. Eng., 1–8.

Shi, N., Tan, L., Yang, C., He, C., Xu, J., Lu, Y., Xu, H., 2021. BacS: a blockchain-based access control scheme in distributed internet of things. Peer-to-Peer Network. Appl. 14 (5), 2585–2599.

Shuaib, M., Alam, S., Nasir, M.S., Alam, M.S., 2021. Immunity credentials using self-sovereign identity for combating COVID-19 pandemic. Mater. Today: Proc. https://doi.org/10.1016/j.matpr.2021.03.096.

Singh, M., Aujla, G.S., Bali, R.S., Vashisht, S., Singh, A., Jindal, A., 2020. Blockchain-enabled secure communication for drone delivery: a case study in COVID-like scenarios. In: Proceedings of the 2nd ACM MobiCom Workshop on Drone Assisted Wireless Communications for 5G and beyond, pp. 25–30.

Song, J., Gu, T., Fang, Z., Feng, X., Ge, Y., Fu, H., Mohapatra, P., 2021. Blockchain meets COVID-19: a framework for contact information sharing and risk notification system. In: 2021 IEEE 18th International Conference on Mobile Ad Hoc and Smart Systems (MASS). IEEE, pp. 269–277.

Sookhak, M., Jabbarpour, M.R., Safa, N.S., Yu, F.R., 2021. Blockchain and smart contract for access control in healthcare: A survey, issues and challenges, and open issues. J. Netw. Comput. Appl. https://doi.org/10.1016/j.jnca.2020.102950.

Staples, M., Chen, S., Falamaki, S., Ponomarev, A., Rimba, P., Tran, A.B., Zhu, J., 2017. Risks and Opportunities for Systems Using Blockchain and Smart Contracts. Data61, CSIRO, Sydney.

Sust, P.P., Solans, O., Fajardo, J.C., Peralta, M.M., Rodenas, P., Gabaldà, J., Piera-Jimenez, J., 2020. Turning the crisis into an opportunity: digital health strategies deployed during the COVID-19 outbreak. JMIR Public Health Surveill. 6 (2), e19106.

Teilen, 2020. SSI in the Age of a Global Pandemic: Covid Credentials Initiative. Validated ID https://www.validatedid.com/post-de/self-sovereign-identity-in-the-age-of-a-global-pandemic-validated-id-joins-the-covid-credentials-initiative. (Accessed 22 January 2021).

Tsoi, K.K., Sung, J.J., Lee, H.W., Yiu, K.K., Fung, H., Wong, S.Y., 2021. The way forward after COVID-19 vaccination: vaccine passports with blockchain to protect personal privacy. BMJ Innov. 7 (2).

Westphal, E., Seitz, H., 2021. digital and decentralized management of patient data in healthcare using blockchain implementations. Front. Blockchain 36.

Wood, G., 2014. Ethereum: a secure decentralised generalised transaction ledger. Ethereum Project Yellow Paper 151 (2014), 1–32.

Xia, Q.I., Sifah, E.B., Asamoah, K.O., Gao, J., Du, X., Guizani, M., 2017. MeDShare: trustless medical data sharing among cloud service providers via blockchain. IEEE Access 5, 14757–14767.

Xu, J., Liu, H., Han, Q., 2021. Blockchain technology and smart contract for civil structural health monitoring system. Computer-Aided Civil Infrastruct. Eng. 36.

Xu, H., Zhang, L., Onireti, O., Fang, Y., Buchanan, W.J., Imran, M.A., 2020. Beeptrace: blockchain-enabled privacy-preserving contact tracing for covid-19 pandemic and beyond. IEEE Internet Things J. 8 (5), 3915–3929.

Yang, G.Z., Nelson, J., Murphy, R.R., Choset, H., Christensen, H., Collins, S., McNutt, M., 2020. Combating COVID-19—the role of robotics in managing public health and infectious diseases. Science. Robotics 5 (40), eabb5589.

Zhang, Y., Kasahara, S., Shen, Y., Jiang, X., Wan, J., 2018a. Smart contract-based access control for the internet of things. IEEE Internet Things J. 6 (2), 1594–1605.

Zhang, P., White, J., Schmidt, D.C., Lenz, G., Rosenbloom, S.T., 2018b. FHIRChain: applying blockchain to securely and scalably share clinical data. Comput. Struct. Biotechnol. J. 16, 267–278.

Zheng, X., Sun, S., Mukkamala, R.R., Vatrapu, R., Ordieres-Meré, J., 2019. Accelerating health data sharing: a solution based on the internet of things and distributed ledger technologies. J. Med. Internet Res. 21 (6), e13583.

Zheng, L., Xiao, C., Chen, F., Xiao, Y., 2020. Design and Research of a Smart Monitoring System for 2019-nCoV Infection-Contact Isolated People Based on Blockchain and Internet of Things Technology. Research Square.

Zou, W., Lo, D., Kochhar, P.S., Le, X.B.D., Xia, X., Feng, Y., Xu, B., 2019. Smart contract development: challenges and opportunities. IEEE Trans. Softw. Eng. 47.

Digital transformation in robotic rehabilitation and smart prosthetics

Meena Gupta[a], Divya Pandey[b], and Prakash Kumar[c]

[a]Amity Institute of Physiotherapy, Amity University, Noida, Uttar Pradesh, India [b]Sai Institute of Paramedical and Allied Sciences, Dehradun, Uttar Pradesh, India [c]Amity Institute of Occupational Therapy, Amity University, Noida, Uttar Pradesh, India

1. Introduction

Physiotherapy is a significant part of recuperation after a neurological injury such as a stroke or spinal cord injury (Langhorne et al., 2011). With the increasing geriatric populace and associated sicknesses, there is an increased demand for physiotherapy services. Utilizing robots to help with treatment is a promising way to deal with increased demand (Krebs and Hogan, 2012).

Research into robotic rehabilitation has increased exponentially during the past two decades and the number of therapeutic robots has increased drastically. Rehabilitation through robotic therapy can assist in high-frequency and intense training for the affected individual (Changa and Kim, 2013). Robotic technology has grown of late, with faster and more potent computers and computational strategies (Esquenazi and Packel, 2012) that have led to accessible robotic rehabilitation technology. The robot can be depicted as a multifunctional, reprogrammable user intended to locomote material, parts, or devices through factor customized movements to achieve a task (Pignolo, 2009). The main benefit of utilizing robot-based technology in rehabilitation intercession is the capacity to convey high-dosage and high-intensity training (Sivan et al., 2011). This feature makes robotic rehabilitation a favorable technology for patient recovery (Stein, 2012).

While there are benefits in utilizing a robot, there are also difficulties. A significant challenge for robots is the need to at minimum copy the abilities of a prepared therapist (Sanguineti et al., 2009). Specialists, for instance, can evaluate every individual's physical limitation and scale their assistance depending on their needs. They can likewise provide resistance to movements when suitable. However, most robots are intended to give consistent help without considering every patient's functional capacity. Subsequently, treatment isn't customized to every patient's need, and the capacity of a robot to induce neural plasticity is conceivably negated.

One way to manage this is to effectively engage the patient in the robotic training process. Scientists have endeavored to accomplish this by fusing control algorithms that require the patient to actively start movements to play out the task (Marchal-Crespo and Reinkensmeyer, 2009). This algorithm allows the user to control the measure of help given by the robot during the training. Hence, by diminishing the guidance given by the robot, the participant can be encouraged to start active movements.

In addition to creating proper control algorithms, one more significant part of treatment is to take advantage of the plasticity in the nervous system. The corticospinal system has the amazing capacity to go through structural (Bezzola et al., 2011) and functional adjustments (Gaser and Schlaug, 2003) in response to motor training (Hanggi et al., 2010). However, repetitive motor movement alone isn't adequate to drive cortical revamping. Additionally, motor training should also include skill acquisition to induce neuronal plasticity, which is indicative of motor recovery (Plautz et al., 2000). While the idea of combining robotic training with motor learning tasks to further develop active participation has been researched in the upper extremities (Norouzi-Gheidari et al., 2012), there is an absence of exploration on this issue in the lower extremities.

2. Prosthetics

In medicine, a prosthesis is an artificial device that replaces a missing body part that might be lost through injury, infection, or a birth defect. Prostheses are expected to reestablish the ordinary elements of the missing body part.

2.1 Myoelectric prosthetics

Recovering the capacity to walk or perform exercises is the main objective in any prosthesis for restoration. From being a passive device, prosthetic innovation has progressed in utilizing dynamic components, such as biosignals for creating a mechanized prosthesis. The significant thought in creating a prosthesis is to utilize the leftover appendage for the function and attachment of the prosthesis. Continuous utilization of a prosthesis aids rehabilitation for amputees.

More than 3 million people live with upper limb amputation around the world, directly altering the quality of life of an individual (Ziegler-Graham et al., 2008). Luckily, there are a few prosthetic arrangements accessible in the market that attempt to reestablish a portion of the function and characteristics of the amputated limb. Usually, the prosthetic gadgets can be body-powered, pneumatically controlled, or electrically powered (Childress, 1985).

The body-controlled gadgets harness energy from muscles to work the link through a connection. The advantages of body-fueled devices are minimal expense while also being more affordable to fix. These gadgets are not cosmetically engaging and are hard to work with body power by certain clients. The electric prosthetic gadgets that use batteries are in huge demand because of their corrective appearance. However, these gadgets are costly, bulky, and not cheap to fix. Nevertheless, there has been a significant leap forward in the activity of electric prosthetic devices. These remotely controlled devices might be operated from pressure, switches, strain measures, myoelectric signals (MES), and electroencephalogram signals. MES has been utilized in different applications, specifically to possibly control assistive devices for amputees as well as orthotic gadgets and exoskeletons to expand the capability of the client.

These prosthetics are exceptionally robust and versatile to evolving circumstances. An illustration of such a shrewd prosthetic is the myoelectric prosthetic hand from Otto Bock HealthCare GmbH (Duderstadt, Germany) (Figs. 1–3).

Artificial hands are cosmetically satisfying yet practically inferior compared to hooks and prehensors. These artificial hands might be controlled utilizing MES, mirroring the aim of the client. The myoelectric prosthetic hand depends on electromyographic (EMG) signals produced in skeletal muscles, which reflect the objective of the client. The EMG signals are utilized to control the prosthetic hand with different translating plans such as relative control, on-off control, recognition of patterns, and postural control. Researchers are endeavoring to decipher additional data from EMG to work on the ability of prosthetic hands. Some researchers are working on EMG signal interfacing strategies to work on dexterity. In any case, prosthetic hand control relies upon the method of detecting the signals as well as translating the intention from the EMG signals (Geethanjali, 2016). A few developed myoelectric hands are:

- **Michelangelo hand**: It includes a thumb that can electronically move into position, and it can perform different grasp capacities.
- **Bebionic 3:** Represents a multiarticulating myoelectric hand that includes 14 grip patterns. This prosthesis is currently available in nine different skin shades. Due to its optimized weight distribution, the hand feels lighter in weight.
- **i-Limb**: The i-limb highlights five independently controlled digits and completely articulated joints. It highlights key holds with accuracy and can be utilized for activities that need power.

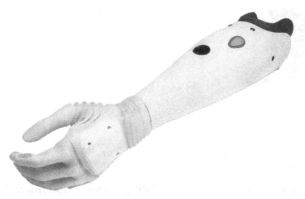

FIG. 1   The Michelangelo hand.

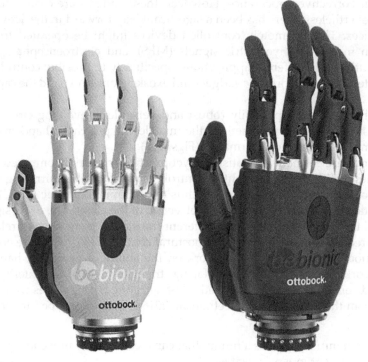

FIG. 2   Bebionic 3.

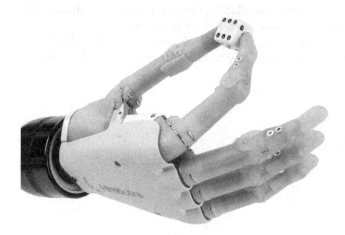

FIG. 3   i- Limb.

2.2 Robotic rehabilitation systems

Robotic exoskeletons have emerged as rehabilitation tools that might improve a few of the current health-related consequences after a brain injury, stroke, SCI, and many more. Every year, 55.9 million individuals experience the ill effects of acquired brain injury, 15 million experience the ill effects of stroke, and up to 500,000 individuals experience the ill effects of SCI. Restricted mobility or loss of motion of some type is often observed in these affected individuals. This can be an overwhelming diagnosis for patients and their families. Robotic exoskeletons are viewed as wearable robotic units managed by a computer to drive an arrangement of engines, pneumatics, levers, or hydraulics to reestablish motion (Gorgey et al., 2018; Miller et al., 2016). Contrasted with previous locomotor training standards, exoskeletons might offer a lot of independence in clinical centers and communities, including shopping centers, neighborhood parks, and cinemas, while increasing the level of physical therapy (Federici et al., 2015).

3. Upper limb exoskeletons

Neurological debilitation such as a stroke alters the functional mobility of survivors and impacts their ability to perform activities of everyday living. Lately, automated restoration for upper appendage recovery has gotten increased attention because it can give high-intensity and repetitive movement treatment. Robotic restoration and the utilization of robotics in rehabilitaion has animated wide concern overall (Zhang and Wang, 2013). To help patients accomplish their functional recovery or to provide remuneration for the loss of motor function, a series of new smart rehabilitation robots has been made fused with prostheses and mechanical auxiliary frameworks (Tan and Wang, 2013; Jamwal et al., 2016).

There are two kinds of robotic rehabilitation systems: one is the end-effector upper limb rehabilitation robot and the other is the exoskeleton rehabilitation robot (Gopura and Kiguchi, 2009). These robots can be utilized to direct patients to finish the assigned recovery task (Fasoli et al., 2003) while giving repetitive and intensive physiotherapy (Riener et al., 2005; Colombo et al., 2015; Norouzi-Gheidari et al., 2012).

3.1 End-effector rehabilitation robot

The end-effector restoration robot structure contains a traditional associating bar and a series robot instrument. In the working state, the robot drives the development of the upper appendage by associating with the patient's arm to achieve rehabilitation training. The robot structure is to some degree independent of the patient, who simply interfaces with the end of the robot.

The upper appendage recuperation robot was first displayed in 1993. Lum et al. fostered a "hand-object-hand" structure used in the upper appendage of hemiplegic stroke patients after a period of recovery (Lum et al., 1992). The bend or extension of the wrist with the help of a drive engine can be accomplished. Starting around 1995, Krebs et al. made MIT rehabilitation robots and broadened their ability. So far, the mechanical bits of the robot have three modules: a graphic module, a wrist module, and a hand module. The robot assists patients with completing the drawing of the elbow, forearm, and wrist joint movement and grasping training. Twenty hemiparetic stroke patients were divided into a benchmark bunch and a preliminary bunch in the clinical preliminary. The last option was expected to execute traction of the shoulder and elbow with a machine. Results showed that patients in the preliminary bundle worked further and quicker (Fugl-Meyer ($P \leq 0.20$) and motor power scores ($P \leq 0.10$), motor status score ($P \leq 0.05$) (Krebs et al., 1998).

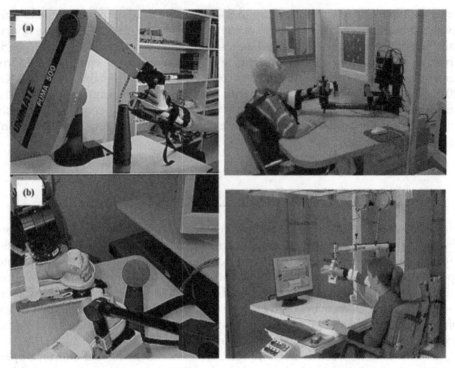

FIG. 4   MIME system A *(left)*, Inmotion2 (the commercial version of MIT-MANUS) [*right (up)*], and GENTLE/s system [*right (down)*].

In 2000, Stanford University made a robot called the MIME upper limb rehabilitation robot. It can help patients with upper limb recuperation training in the mirror representation of the contralateral developments (Kahn et al., 2006). Two subacute stroke patients recognized the treatment during three weeks. The Fugl-Meyer, Box and Block Tests, and the Jebsen-Taylor Test showed enhancement of the patients' upper appendage, especially the finger and hand (Daud et al., 2011).

An upper limb rehabilitation robot named GENTLES was created by Reading University in 2002. It was the opportunity to apply virtual reality development in a recovery robot. Patients can complete the training independently by gravity compensation work and visual feedback (Amirabdollahian et al., 2007). Eight subjects partook in the analysis; it was announced that this system is more sensible for those stroke patients with specific athletic limits (Lam et al., 2008) (Fig. 4).

3.2 Exoskeleton rehabilitation robot

In end effector upper appendage, robots can manage perplexing direction during training sessions, but its difficult to manage it in another theraputic sessions. As such, there is another kind of upper appendage recovery robot framework accessible—the exoskeleton of the upper appendage restoration robot. An exoskeleton robot is a wearable gadget that joins mechanical power devices and intelligent control. Essentially, exoskeleton robots can be separated into upper-appendage exoskeleton robots, lower-appendage exoskeleton robots, whole-body exoskeleton robots, and a wide scope of joint-adjustment robots. The exoskeleton recovery robot gives power compensation, body insurance, and support. It has joined sensor and control information, which can be put together with the patient to complete the auxiliary training of body movements.

The University of Arizona made an artificial 4-DOF robot driven by the pneumatic muscle (PM) and a 5-DOF upper appendage rehabilitation robot called RUPERT. A development driven by the pneumatic muscle is more versatile because of comparable muscle limit and development characteristics (Sugar et al., 2007). The altered Wolf Motor Test and the Fugl-Meyer upper motor test were utilized. Eight sound subjects and two stroke patients took part in this assessment (Balasubramanian et al., 2008). Four sound subjects worked on both the basic and complex exercises, and two patients further fostered the limb function (Zhang et al., 2011).

The University of California along with the University of California at Irvine created the 5-DOF T-WREX (training Wilmington robotic exoskeleton) (Sanchez et al., 2004). It is sensible for active rehabilitation training and development parameter assessment for patients without drive and 5-DOF Pneu-WREX upper appendage restoration robot structures to train the shoulder and elbow (Sanchez et al., 2005; Rosen et al., 2005) which chips away at the pneumatic drive, which is further suitable for patients with passive rehabilitation training. Five stroke patients took an interest in this study which shows that gravity balance training further developed reaching capacity to the contralateral target however not to the ipsilateral target and further developed the vertical reaching.

Another great example would be MAHI Exo II, a 5-DOF forearm rehabilitation robot developed by Rice University (Pehlivan et al., 2011). The primary portion of the robot was a 3-DOF parallel mechanism that can understand wrist bending, stretch, adduction, extension, and complete arm stretch. Another part was a 2-DOF serial mechanism that can attain forearm rotation, elbow bend, and stretch. In synopsis, the exoskeleton robot can understand more precisely the assist motion, and the end-effector rehabilitation robot has better execution on feedback and assessment.

4. Lower limb exoskeletons

Debilitated mobility is the main issue in recovery (King et al., 2018) and can be brought about by arthritis, stroke, neural injury from traumatic brain injury, Parkinson's, and spinal cord injuries (SCIs) (O'Sullivan and Schmitz, 2016). SCIs are pervasive worldwide; they are devastating to people and their families, and are costly to treat and rehabilitate (Lee et al., 2012). Stroke-related paralysis is the most common adult physical impairment, requiring intensive rehabilitation to recover functionality (Langhorne et al., 2009). Mobility debilitation can cause secondary degenerative changes in muscles and neural processes (Miller et al., 2016) that can lead to muscle atrophy and pain as well as psychological and behavioral issues such as depression (Louie et al., 2015).

Improving lower limb function and the ability to walk are significant needs for people who have lost locomotor capacity and are typically focused on physical therapy.

As of late, robotic exoskeletons have arisen as a gait rehabilitation choice that can reduce a portion of the physical demands of lower extremity rehabilitation. There are a few diverse rehabilitation exoskeletons accessible from various companies (Holanda et al., 2017). While not indistinguishable, their overall design comprises an externally mechanized structure fitted over and around weak or paralyzed limbs to assist with standing, walking, and other day-to-day activities. The below mentioned are some smart lower limb exoskeletons accessible on the market.

4.1 HAL

The hybrid assistive limb (HAL) was developed by Tsukuba University and Cyberdyne in Japan. It is a wearable exoskeleton that is designed for different applications, including rehabilitation, heavy labor support, rescue support, etc. It comes in different versions: full body, lower body, and one leg (Kawamoto et al., 2009). The single-leg version is specially designed for individuals with hemiplegia. A newer version of this called HAL5 (full body) is designed for paraplegic individuals. HAL received a global ISO safety certificate in February 2013, making it the first exoskeleton to achieve this goal. Subsequent safety-certified models were released in 2014 as well (ISO 13482, 2014). It is power operated and has power independence to work for roughly 2.5h. Hip and knee joint actuators are based on DC servo engines and harmonic drive gears while the ankle joint is controlled passively. The actual mechanism of working is based on surface electromyography (sEMG) in which surface electrodes in contact with the skin provide the system with the wearer's intended movements. Power assistance that is provided by HAL sometimes makes it difficult for people with severe hemiplegia to perform tasks utilizing their own muscles, which often leads to instability as well as a decrease in stride length and walking speed (Figs. 5 and 6).

4.2 ReWALK

ReWalk is a wearable mechanized medical suit from Argo Medical Technologies (Talaty et al., 2013) that can be utilized for therapeutic exercises. ReWALK comprises a light wearable brace support suite that integrates with the DC motor at the hip and knees. It also has rechargeable batteries, sensors, and a computer-controlled system. It can control knee and hip movements in the sagittal plane and is customized for every patient (Mohammed and Amirat, 2009). In this gadget, variations in the client's center of gravity are recognized to start and keep up with walking. The client additionally has a controller put in his/her arm, similar

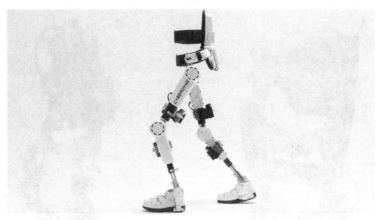

FIG. 5   Hybrid assistive limb (HAL).

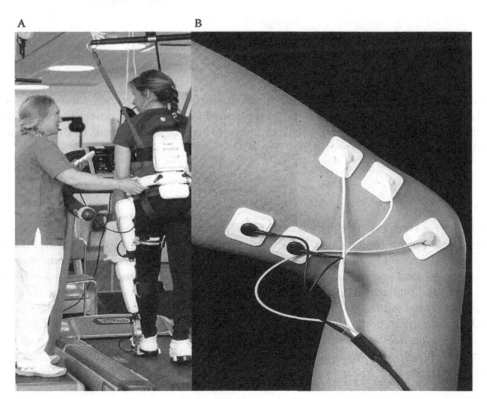

FIG. 6   HAL. (A) Illustration of HAL training, (B) surface electromyography to capture the wearer's voluntary muscle activation.

B. Enabling technologies for digital transformation (an example-oriented approach)

A    B

FIG. 7   ReWalk.

to a watch. With this interface, it is conceivable to begin various errands, for example, sit to stand, walking, or climbing steps. ReWalk is planned for people with lower limb disabilities that have endured wounds in the spinal cord. It can't keep balance control, so the client ought to continuously be upheld by crutches (Fig. 7).

4.3 eLEGS

Ekso Bionics (formerly Berkeley Bionics) is a United States organization that initially created exoskeletons for military use. In October 2010, they revealed the exoskeleton lower extremity gait system (eLEGS), which was renamed Ekso in 2011. The Ekso weighs around 20 kg and has the greatest speed of 3.2 km/h with a battery life of 6 h. Activities such as sit to stand, stand to sit, and walking in a straight line can be executed by EKso. It is presently being advanced to become lighter and more versatile. Following clinical preliminaries, the US Food and Drug Administration approved its use in 2012. The gadget can be instructed by a user interface that can handle the gadget bit by bit. Like the ReWalk framework, Ekso also needs crutches for the user to help adjust the weight because of the imbalance during the initial jerk in the movements (Fig. 8).

FIG. 8   e-LEGS.

5. Discussion

Throughout the last decade, the presentation of robotic advancements into rehabilitation settings has advanced from idea to reality. The studies show the productivity and benefits of robotic rehabilitation for evaluating and treating motor weaknesses in both the upper and lower limbs (Lum et al., 2002). A few studies have concurred that the high-intensity repetitive movements comprise a significant contribution to the viability of robot-aided treatment (Masiero et al., 2009). However, there is a huge hole between the results of rehabilitation robots and individuals' assumptions (Hashimoto et al., 2006). Equipment to work on the quality of physiotherapy is scant (Campolo et al., 2009). One issue is the trouble of personalization. It is alluring to give robotic help to patients in such a way that the subsequent interaction between the robot and the patient is smoother. The degree of help ought to be tuned as per the patients' capacities (Johnson et al., 2007). Each patient has his/her particular capacities, functional requirements, and interests. An accentuation on more independent utilization of robotic treatment frameworks makes the personalization of the human-innovation interface vital (Mokhtari et al., 2006).

There are two critical parts of the customized interfaces: the physical interface (e.g., the gadget itself, its actual settings, and the scope of activity of a gadget comparative with the client's torso) and the communication interface. The more noteworthy research challenge is how to customize and alter the focus of therapeutic intervention, particularly when a patient exhibits some improvement. This proposes the significance of a training protocol that is often varied, in terms of utilization of both the full "capacity" work area and the kinds of tasks performed inside the work area. The design of robots devoted to individuals with inabilities requires clients' suggestions in all areas of product development: design, solution, prototyping the framework, selection of client' interfaces, and testing with clients in genuine conditions. Prior to any framework, it is important to comprehend and meet the requirements of the disabled clients (Mahoney, 1998)

Cost is one more significant issue in promoting rehabilitation robots. The price/performance proportion is dissatisfactory because of the high expense along with a low advantage for patients and clinic set-ups. As a synopsis, robotic rehabilitation is entering the market gradually but is yet to be seen as future technology. The basic issues are the trouble of personalization and high cost. For low-to-middle pay classes, just 5%–15% of individuals who need assistive gadgets and technologies can access these advances.

There is a shortage of staff prepared to deal with the provision of such gadgets and advancements. Moreover, research and development projects in robotic rehabilitation are viewed critically in terms of their results and outcome. Nonetheless, the innovative work on rehabilitation robots is arising because of the fact that the expense of preventing individuals with incapacities from a functioning part in community life is high and must be borne by society, especially the people who assume the weight of care. This rejection regularly prompts misfortunes in human potential and efficiency (Bi et al., 2010)

6. Conclusion

Many industrialized nations are experiencing restricted assets for medical care and an increased population of disabled and geriatric individuals. A practical way to address this issue is to apply further developed and reliable robotic advances for the medical care industry. Existing assistive advancements are unacceptable in addressing individual requirements and satisfying the necessary capacities at a reasonable expense. Any advancement to work on the ability of personalization or reducing the proportion of cost and execution will invigorate the use of mechanical advances in medical care significantly. Reconfigurability and modularization can be the answers to address these issues; they can advance the utilization of robotic innovations for additional functionalities and designated patients. A reconfigurable modular robot comprises a bunch of modules. Various types and numbers of modules can be conveyed and these modules can be assembled in various set-ups. Subsequently, a similar framework can produce numerous set-ups for meeting the various needs of patients. Thus, a similar arrangement of modules can be used in numerous applications; this prompts a higher pace of utilization. The procedures for reconfigurable secluded robots can be certainly expanded and utilized in designing other medical robots.

References

Amirabdollahian, F., Loureiro, R., Gradwell, E., Collin, C., Harwin, W., Johnson, G., 2007. Multivariate analysis of the Fugl-Meyer outcome measures assessing the effectiveness of GENTLE/S robot-mediated stroke therapy. J. Neuroeng. Rehabil. 4 (1), 4. https://doi.org/10.1186/1743-0003-4-4.

Balasubramanian, S., Wei, R., Perez, M., et al., 2008. RUPERT: an exoskeleton robot for assisting rehabilitation of arm functions. In: Virtual Rehabilitation, pp. 163–167. Vancouver BC, Canada.

Bezzola, L., Merillat, S., Gaser, C., Jancke, L., 2011. Training-induced neural plasticity in golf novices. J. Neurosci. 31, 12444–12448.

Bi, Z.M., Lin, Y., Zhang, W.J., 2010. The general architecture of adaptive robotic systems for manufacturing applications. Robot. Comput. Integr. Manuf. 26 (5), 461–470.

Campolo, D., Accoto, D., Formica, D., Guglielmelli, E., 2009. Intrinsic constraints of neural origin: assessment and application to rehabilitation robotics. IEEE Trans. Robot. 25 (3), 492–501.

Changa, W.H., Kim, Y.-H., 2013. Robot-assisted therapy in stroke rehabilitation. J. Stroke 15 (3), 174–181.

Childress, D.S., 1985. Historical aspects of powered limb prosthesis. Clin. Prosthet. Orthot. 9 (1), 2–13.

Colombo, R., Sterpi, I., Mazzone, A., Delconte, C., Pisano, F., 2015. Improving proprioceptive deficits after stroke through robot-assisted training of the upper limb: a pilot case report study. Neurocase 22 (2), 191–200. https://doi.org/10.1080/13554794.2015.1109667.

Daud, O.A., Oboe, R., Agostini, M., Turolla, A., 2011. Performance evaluation of a VR-based hand and finger rehabilitation program. In: 2011 IEEE International Symposium on Industrial Electronics, pp. 934–939. Gdansk, Poland.

Esquenazi, A., Packel, A., 2012. Robotic-assisted gait training and restoration. Am. J. Phys. Med. Rehabil. 91, S217–S227. quiz S228-231.

Fasoli, S.E., Krebs, H.I., Stein, J., Frontera, W.R., Hogan, N., 2003. Effects of robotic therapy on motor impairment and recovery in chronic stroke. Arch. Phys. Med. Rehabil. 84 (4), 477–482. https://doi.org/10.1053/apmr.2003.50110.

Federici, S., Meloni, F., Bracalenti, M., De Filippis, M.L., 2015. The effectiveness of powered, active lower limb exoskeletons in neurorehabilitation: a systematic review. NeuroRehabilitation 37, 321–340.

Gaser, C., Schlaug, G., 2003. Brain structures differ between musicians and non-musicians. J. Neurosci. 23, 9240–9245.

Geethanjali, P., 2016. Myoelectric control of prosthetic hands: state-of-the-art review. Med. Devices Evid. Res. 9, 247–255.

Gopura, R.A.R.C., Kiguchi, K., 2009. Mechanical designs of active upper-limb exoskeleton robots: state-of-the-art and design difficulties. In: 2009 IEEE International Conference on Rehabilitation Robotics, Kyoto, Japan, pp. 178–187.

Gorgey, A., Sumrell, R., Goetz, L., 2018. Exoskeletal assisted rehabilitation after spinal cord injury. In: Atlas of Orthoses and Assistive Devices, fifth ed. Elsevier, Canada, pp. 440–447.

Hanggi, J., Koeneke, S., Bezzola, L., Jancke, L., 2010. Structural neuroplasticity in the sensorimotor network of professional female ballet dancers. Hum. Brain Mapp. 31, 1196–1206.

Hashimoto, Y., Komada, S., Hirai, J., 2006. Development of a biofeedback therapeutic exercise supporting manipulator for lower limbs. In: Proceedings of the IEEE International Conference on Industrial Technology (ICIT '06), pp. 352–357.

Holanda, L.J., Silva, P.M., Amorim, T.C., Lacerda, M.O., Simão, C.R., Morya, E., 2017. Robotic assisted gait as a tool for rehabilitation of individuals with spinal cord injury: a systematic review. J. Neuroeng. Rehabil. 14 (1), 126.

ISO 13482, 2014. Robots and Robotic Devices: Safety Requirements for Personal Care Robots.

Jamwal, P.K., Hussain, S., Ghayesh, M.H., Rogozina, S.V., 2016. Impedance control of an intrinsically compliant parallel ankle rehabilitation robot. IEEE Trans. Ind. Electron. 63 (6), 3638–3647. https://doi.org/10.1109/TIE.2016.2521600.

Johnson, M.J., Feng, X., Johnson, L.M., Winters, J.M., 2007. Potential of a suite of robot/computer-assisted motivating systems for personalized, home-based, stroke rehabilitation. J. NeuroEng. Rehabil. 4, article 6.

Kahn, L.E., Zygman, M.L., Rymer, W.Z., Reinkensmeyer, D.J., 2006. Robot-assisted reaching exercise promotes arm movement recovery in chronic hemiparetic stroke: a randomized controlled pilot study. J. Neuroeng. Rehabil. 3 (1), 12–290. https://doi.org/10.1186/1743-0003-3-12.

Kawamoto, H., Hayashi, T., Sakurai, T., Eguchi, K., Sankai, Y., 2009. Development of single leg version of HAL for hemiplegia. In: 31st Annual International Conference of the IEEE EMBS Minneapolis, Minnesota, USA, pp. 2–6.

King, L.K., Kendzerska, T., Waugh, E.J., Hawker, G.A., 2018. Impact of osteoarthritis on difficulty walking: a population-based study. Arthritis Care Res. 70 (1), 71–79.

Krebs, H.I., Hogan, N., 2012. Robotic therapy: the tipping point. Am. J. Phys. Med. Rehabil. 91, S290–S297.

B. Enabling technologies for digital transformation (an example-oriented approach)

Krebs, H.I., Hogan, N., Aisen, M.L., Volpe, B.T., 1998. Robot-aided neurorehabilitation. IEEE Trans. Rehabil. Eng. 6 (1), 75–87. https://doi.org/10.1109/86.662623.

Lam, P., Hebert, D., Boger, J., et al., 2008. A haptic-robotic platform for upper-limb reaching stroke therapy: preliminary design and evaluation results. J. Neuroeng. Rehabil. 5 (1), 15. https://doi.org/10.1186/1743-0003-5-15.

Langhorne, P., Bernhardt, J., Kwakkel, G., 2011. Stroke rehabilitation. Lancet 377, 1693–1702.

Langhorne, P., Coupar, F., Pollock, A., 2009. Motor recovery after stroke: a systematic review. Lancet Neurol. 8 (8), 741–754.

Lee, H., Lee, B., Kim, W., Gil, M., Han, J., Han, C., 2012. Human-robot cooperative control based on pHRI (Physical Human-Robot Interaction) of exoskeleton robot for a human upper extremity. Int. J. Precis. Eng. Manuf. 13 (6), 985–992.

Louie, D.R., Eng, J.J., Lam, T., 2015. Gait speed using powered robotic exoskeletons after spinal cord injury: a systematic review and correlational study. J. Neuroeng. Rehabil. 12 (1), 82.

Lum, P.S., Burgar, C.G., Shor, P.C., Majmundar, M., van der Loos, M., 2002. Robot-assisted movement training compared with conventional therapy techniques for the rehabilitation of upperlimb motor function after stroke. Arch. Phys. Med. Rehabil. 83 (7), 952–959.

Lum, P.S., Reinkensmeyer, D.J., Lehman, S.L., Li, P.Y., Stark, L.W., 1992. Feedforward stabilization in a bimanual unloading task. Exp. Brain Res. 89 (1), 172–180.

Mahoney, R.M., 1998. Robotic products for rehabilitation: status and strategy. In: Proceedings of the International Conference of Rehabilitation Robotics. Bath University, Bath, UK, pp. 12–17.

Marchal-Crespo, L., Reinkensmeyer, D.J., 2009. Review of control strategies for robotic movement training after neurologic injury. J. Neuroeng. Rehabil. 6, 20.

Masiero, S., Carraro, E., Ferraro, C., Gallina, P., Rossi, A., Rosati, G., 2009. Upper limb rehabilitation robotics after stroke: a perspective from the University of Padua, Italy. J. Rehabil. Med. 41 (12), 981–985.

Miller, L.E., Zimmermann, A.K., Herbert, W.G., 2016. Clinical effectiveness and safety of powered exoskeleton-assisted walking in patients with spinal cord injury: systematic review with meta-analysis. Med. Devices Evid. Res. 9, 455–466.

Mohammed, S., Amirat, Y., 2009. Towards intelligent lower limb wearable robots: challenges and perspectives—state of the art. In: Proceedings of the 2008 IEEE International Conference on Robotics and Biomimetics, pp. 312–317.

Mokhtari, M., Feki, M.A., Abdulrazak, B., Grandjean, B., 2006. 3 Toward a human-friendly user interface to control an assistive robot in the context of smart homes. In: Advances in Rehabilitation Robotics, vol. 306 of Lecture Notes in Control and Information Science. Springer, Berlin, Germany, pp. 47–56.

Norouzi-Gheidari, N., Archambault, P.S., Fung, J., 2012. Effects of robot-assisted therapy on stroke rehabilitation in upper limbs: systematic review and meta-analysis of the literature. J. Rehabil. Res. Dev. 49, 479–496.

O'Sullivan, S.B., Schmitz, T.J., 2016. Improving Functional Outcomes in Physical Rehabilitation, second ed. F.A. Davis Company, Philadelphia, PA.

Pehlivan, A.U., Celik, O., O'Malley, M.K., 2011. Mechanical design of a distal arm exoskeleton for stroke and spinal cord injury rehabilitation. In: 2011 IEEE International Conference on Rehabilitation Robotics, Zurich, Switzerland, pp. 1–5.

Pignolo, L., 2009. Robotics in neuro-rehabilitation. J. Rehabil. Med. 41, 955–960.

Plautz, E.J., Milliken, G.W., Nudo, R.J., 2000. Effects of repetitive motor training on movement representations in adult squirrel monkeys: role of use versus learning. Neurobiol. Learn. Mem. 74, 27–55.

Riener, R., Nef, T., Colombo, G., 2005. Robot-aided neurorehabilitation of the upper extremities. Med. Biol. Eng. Comput. 43 (1), 2–10. https://doi.org/10.1007/BF02345116.

Rosen, J., Perry, J.C., Manning, N., Burns, S., Hannaford, B., 2005. The human arm kinematics and dynamics during daily activities—toward a 7 DOF upper limb powered exoskeleton. In: ICAR '05. Proceedings, 12th International Conference on Advanced Robotics, 2005, Seattle, WA, USA, pp. 532–539.

Sanchez, R., Reinkensmeyer, D., Shah, P., et al., 2004. Monitoring functional arm movement for home-based therapy after stroke. In: The 26th Annual International Conference of the IEEE Engineering in Medicine and Biology Society, September 2004, San Francisco, CA, USA, pp. 4787–4790.

Sanchez, R.J., Wolbrecht, E., Smith, R., et al., 2005. A pneumatic robot for re-training arm movement after stroke: rationale and mechanical design. In: 9th International Conference on Rehabilitation Robotics, 2005. ICORR 2005, Chicago, IL, USA, pp. 500–504.

Sanguineti, V., Casadio, M., Vergaro, E., Squeri, V., Giannoni, P., et al., 2009. Robot therapy for stroke survivors: proprioceptive training and regulation of assistance. Stud. Health Technol. Inform. 145, 126–142.

Sivan, M., O'Connor, R.J., Makower, S., Levesley, M., Bhakta, B., 2011. Systematic review of outcome measures used in the evaluation of robot-assisted upper limb exercise in stroke. J. Rehabil. Med. 43, 181–189.

Stein, J., 2012. Robotics in rehabilitation: technology as destiny. Am. J. Phys. Med. Rehabil. 91, S199–S203.

Sugar, T.G., He, J., Koeneman, E.J., et al., 2007. Design and control of RUPERT: a device for robotic upper extremity repetitive therapy. IEEE Trans. Neural Syst. Rehabil. Eng. 15 (3), 336–346. https://doi.org/10.1109/TNSRE.2007.903903.

Talaty, M., Esquenazi, A., Briceño, J.E., 2013. Differentiating ability in users of the ReWalk powered exoskeleton. In: IEEE International Conference on Rehabilitation Robotics.

Tan, M., Wang, S., 2013. Research progress on robotics. Acta Automat. Sin. 39 (7), 1119–1128.

Zhang, H., Austin, H., Buchanan, S., Herman, R., Koeneman, J., He, J., 2011. Feasibility study of robot-assisted stroke rehabilitation at home using RUPERT. In: The 2011 IEEE/ICME International Conference on Complex Medical Engineering, Harbin Heilongjiang, China, pp. 604–609.

Zhang, X.Y., Wang, K.X., 2013. Robot assisted rehabilitation technology, robotics and intelligent devices. Chin. J. Rehabil. 28 (4), 246–248.

Ziegler-Graham, K., MacKenzie, E.J., Ephraim, P.L., Travison, T.G., Brookmeyer, R., 2008. Estimating the prevalence of limb loss in the United States: 2005 to 2050. Arch. Phys. Med. Rehabil. 89 (3), 422–429.

Further reading

Jancke, L., Koeneke, S., Hoppe, A., Rominger, C., Hanggi, J., 2009. The architecture of the golfer's brain. PLoS One 4, e4785.

Ensemble deep neural models for automated abnormality detection and classification in precision care applications

K. Karthik, Veena Mayya, and S. Sowmya Kamath

Healthcare Analytics and Language Engineering Lab, Department of Information Technology, National Institute of Technology Karnataka, Surathkal, Mangaluru, India

1. Introduction

The latest developments of Data Science and Artificial Intelligence allow a new generation of value-based healthcare (Spruit and Lytras, 2018; Lytras et al., 2021a, b). Various symptoms of COVID-19 infection include fever, cough, and respiratory illnesses such as shortness of breath and influenza. In some serious cases, the viral infection may lead to difficulty in breathing, pneumonia, loss of taste or smell, chest pain, multiorgan failure, and death (WHO, 2020). Predominant screening methods used for detecting COVID-19 include the reverse transcription polymerase chain reaction test (RT-PCR), which utilizes samples collected from the patients' upper respiratory tract. The rapidly increasing rate of COVID-19 cases caused a shortage of ventilators and testing kits, due to which health systems saw an unprecedented collapses even in countries with access to very advanced healthcare facilities. In view of these prevalent challenges, several alternatives were explored, the most economical among these being X-ray scans. In hospitals, trained radiologists analyze and diagnose diseases using a wide variety of scanning modalities on a daily basis. X-ray is thus one of the most commonly used, accessible, and widely adopted diagnostic imaging modalities. Chest X-rays (CXR) are already commonly used for the rapid examination of lung abnormalities such as the onset of chronic pulmonary disease, lung cancer, pneumonia, etc.

Digital Transformation in Healthcare in Post-COVID-19 Times
https://doi.org/10.1016/B978-0-323-98353-2.00014-9

During the ongoing COVID-19 pandemic, radiography images have been used extensively as inexpensive tools for prescreening and detecting COVID symptoms through lung cavity examination. Recently, researchers have explored the utility of chest radiography images for hastening the diagnosis of COVID-19 (Mayya et al., 2021; Narin et al., 2021). Several works revealed that the occurrence of dominant visual markers such as ground-glass opacities in the lung cavity helped identify the onset of infection (Fang et al., 2020). The time gap from initial symptoms to the imaging procedure is a significant factor affecting the reliability of X-ray findings. In the first 3 days when symptoms such as coughing and fever are observed, initially the infection is not observed on X-rays, but it is most visible after 10–12 days (Osman et al., 2021). Several research studies reported that the radiology-based evaluation can be an effective tool in the detection and screening of COVID-19 cases.

Musculoskeletal disorders are often debilitating and affect a person's daily activities to a large extent. As per statistics published by the US Bone and Joint Initiative (BMUS, 2014), as many as 1.7 billion people worldwide suffer from work- and stress-related musculoskeletal conditions. It is also reported that the occurrence of chronic musculoskeletal conditions in adults is significantly more than that of coronary and heart conditions, and double that of chronic respiratory conditions, which are ranked as the second and third most common health conditions in the world. Hairline fractures, carpal tunnel syndrome, and ligament tears are some examples of such disorders, the potential causes of which could be work-related accidents, workplace fatigue, sport injuries, etc. The daily routines of busy professionals, such as those who do repetitive tasks on computers (banking, software professionals, etc.), have led to increasing occurrence of neck and lower back pain. Researchers who studied occupation-related cases of such professionals observed that physical and individual factors affected the health of the worker's musculoskeletal system over time (Karthik and Kamath, 2022; Silvian et al., 2011; Wærsted et al., 2010).

Abnormalities that exist in medical images can be characterized and detected in various ways. A traditional and common categorization is using a shape model. Certain abnormalities have unique geometric shapes, and their sizes are helpful for the detection of disease staging and prognosis. Considering an example, some tumor regions are round in shape and tend to grow in size. In such cases, the detection of abnormalities can be considered a task with a particular pattern in the medical image. These patterns can be from a collection of geometric shapes such as corners, edges, lines, circles, contours, etc. These shapes are detected by means of a specific method or technique. For example, lines can be detected by various edge detection techniques. Further, detecting the abnormalities can be improved by integrating prior knowledge of abnormalities, such as the location, size, or orientation. One such work (Karthik and Kamath, 2021) that integrates these methods to automatically detect the abnormality region for a quick diagnosis has shown its potential for accurate diagnosis.

The application of deep learning in areas such as computer vision, natural language processing, bioinformatics, and medical image analysis has seen widespread adoption (Ching et al., 2018). In particular, deep convolutional neural networks (CNNs) have shown their power and proved their suitability on large-scale labeled datasets due to their powerful predictive capabilities. In this chapter, we propose an adaptable deep neural model for multiple clinical tasks such as classifying COVID-19 versus normal/no findings in CXR and the automated, quick detection of abnormalities from upper extremity musculoskeletal images. The key contributions of this research include:

1. Designing a deep neural model adaptable for multiple clinical tasks such as disease classification and abnormality detection from radiography images.

2. Designing a boundary detection algorithm for identifying the region of interest for facilitating anomaly detection from radiography images.

3. Experimental evaluation of the performance of the proposed approaches for two different clinical tasks and comparison with existing works.

The rest of this chapter is detailed as follows: Section 2 discusses existing works in the area of interest. In Section 3, we present a detailed discussion of the proposed methods for medical image analysis. Experimental results and discussions are presented in Section 5, followed by conclusion and future work.

2. Related work

Extensive research efforts are evidenced in demonstrating different approaches for COVID-19 detection from CXR images (Narin et al., 2021; Wang et al., 2020). Research based on implicit feature learning-based COVID classification (Shen et al., 2017), CXR image classification for detecting pneumonia (Rajpurkar et al., 2017), etc. has been proposed. Alhudhaif et al. (2021) used transfer learning techniques with three different CNN pretrained models trained on a small dataset with significant data imbalance for COVID-19 classification. Jain et al. (2021) also used transfer learning techniques with Inception V3, Xception, and ResNeXt models for multiclass classification of X-ray images, during which the Xception model showed the best accuracy of about 97.97%. Shakarami et al. (2021) used CNNs to categorize the X-ray images into COVID-19 and non-COVID-19 groups for the classification and retrieval of images, which helps physicians in the further analysis of the most similar cases. Sait et al. (2021) proposed CovScanNet for classifying breathing sound spectrograms and X-ray images. The breathing sound is also trained with the help of a spectrogram by filtering the wheezing sounds in detection along with the X-ray images. An accuracy of 80% in COVID-19 detection from the curated dataset was reported.

Efforts are in progress for developing large, open, and standard repositories for facilitating medical image management (Rajpurkar, 2017). The musculoskeletal radiographs (MURA) dataset made available by Rajpurkar (2017) consists of an extensive collection of musculoskeletal studies of upper extremities such as the hand, wrist, elbow, finger, forearm, humerus, and shoulder radiography classes. Automatically deciding if a radiography study is normal or abnormal is a task that can significantly reduce the burden of radiologists, who handle a large volume and variety of radiological images each day. Rajpurkar (2017) developed a deep 169-layer CNN for predicting the abnormality of musculoskeletal studies. The model predicts a study as abnormal if the prediction rate was found to be greater than 0.5. Here, the variable-sized images were scaled to 320×320 and the network weights were set according to the pretrained ImageNet model with a learning rate of 0.0001 and a minibatch size of 8. Their results performed best for wrist and humerus classes using Cohen's kappa statistic with a 95% confidence interval compared with the three best radiologists.

Tataru et al. (2017) created a pipeline of three neural network architectures, such as GoogLeNet, InceptionNet, and ResNet, to prove its performance on the classification task which used more than 4,00,000 CXR, labeled as normal, abnormal, and extremely high-risk cases. The original image sizes were downsampled to 512×512 and 224×224 from 3000×3000 pixels before input to the neural net. Among the three, GoogleNet performed the best for the CXR image classification task. Lai and Deng (2018) proposed a medical image classification model for two types of datasets, HIS2828 and ISIC2017. They developed a deep learning model called coding network with multilayer perceptron (CNMP) that integrates high-level features extracted from CNN and some traditional features such as texture-based and color moment. Six types of classification algorithms were used to check the classification accuracy and CNMP performed well with an accuracy of 90.1% for ISIC2017 and 90.2% for HIS2828 datasets compared to other classifiers.

Gale et al. (2017) proposed a hip fracture detection approach based on deep neural networks on a small annotated dataset, which includes 3354 images with 348 fractures, and reported good results. Rahmany et al. (2019) used a multistep approach using local binary patterns (LBP) and Fourier descriptors, where the first phase identifies the initial aneurysm using LBP and in the second phase, the false positive cases are identified using Fourier shape descriptors. Their results showed the excellent performance of the approach, with reference to both manual and automated detectors. The computerized image analysis (IA) detection framework correctly detected all aneurysms with low false-negative and false-positive rates.

Hajabdollahi et al. (2018) proposed a method to detect multiple abnormalities in endoscopic and dermoscopic images of skin lesions using a bifurcated CNN, which was used for performing both classification and segmentation simultaneously on multiple abnormalities. The neural network is trained initially by each abnormality separately; later all the abnormalities were combined for training the network. For identifying pulmonary and pleural abnormalities in CXR, Karargyris et al. (2016) developed a feature-based method to classify X-ray images into tuberculosis (TB) and non-TB cases. Shape features were used to identify the dimension and texture features were used to represent the image level aspects. This approach was tested on the Shenzhen and Japanese Society of Radiological Technology (JSRT) datasets, and achieved an AUC improvement of 2.4% over other works. ChexNet (Rajpurkar et al., 2017) is a 121-layer CNN used for the detection of pneumonia from CXR. The model was trained on more than 100,000 CXR images of diseases atelectasis, cardiomegaly, pneumothorax, emphysema, pleural thickening, hernia, and others and the model's prediction accuracy exceeded radiologist's performance. An automatic feature learning and classifier building using CNNs for abnormality detection in mammogram images was proposed by Xi et al. (2018). Experimental results showed that VGGNet's accuracy of 92.53% was the best in class.

Based on the extensive review, several gaps in existing works were identified. Most models focus on identifying COVID-19 cases with various techniques using radiography/CT images and classification of pneumonia among normal cases. Our work encompasses the development of a CNN-based model for automated abnormality classification and detection. Our objective is to demonstrate the suitability of the models across specific clinical tasks, so we consider two diverse datasets for our experiments. One is a curated CXR dataset to demonstrate disease classification. The other is a dataset containing musculoskeletal studies of upper extremity radiography images, for which we aim to demonstrate the suitability of the proposed models for abnormality detection and visualization. The proposed model was trained on the two different datasets after scaling the images as required by the input layer of the adapted CNN architecture. Once the model is trained with a particular study, its testing study set of images is classified using the neural network classifier and the classification accuracy is observed. Further, the abnormalities present in the images are identified using our developed abnormal region detection algorithm and class activation maps. The performance of the model is evaluated with standard evaluation metrics such as accuracy, sensitivity, specificity, and kappa score. The proposed approach is detailed in Section 3.

3. Proposed approach

The series of tasks involved in the proposed approach are illustrated in Fig. 1. We adopted different neural network architectures as an alternative to the traditional supervised learning methods. Two different CNN approaches are used, known for their effective performance in optimized training processes. To visualize the learnt features that contributed most to the prediction, a class activation map (Grad-CAM) model, Selvaraju et al. (2017) was incorporated, which highlights the regions on the input test CXR images. We also developed an abnormal region detection algorithm for identifying the abnormalities present in the musculoskeletal images. The region that gets highlighted is obtained from the neural network classification model using the learned feature vectors performing the prediction on the abnormal region, thus enabling evidence-based diagnosis. To demonstrate the adaptability of the proposed approach for multiple clinical tasks, we considered two different diagnosis tasks—disease classification and abnormality detection. Hence, the model's output layer consists of a single neuron with sigmoid activation, for providing the approximate probability that an input image is abnormal.

The proposed deep neural model, built on the DenseNet (Huang et al., 2017) and AlexNet (Krizhevsky et al., 2012), is designed to classify the CXR and musculoskeletal radiograph images into normal or abnormal classes. The first layer of the proposed model, the input layer, is comprised of three dimensions: *width*, *height*, and *depth*. Hence, the images are rescaled to a size of 224 × 224 and 227 × 227 dimensions and then fed to the input layer of the neural network. The deep CNN model is composed of convolutional layers, after which three fully connected layers are used as in a standard multilayer neural network. Each convolutional

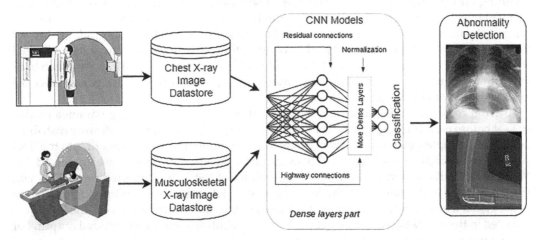

FIG. 1   Proposed approach for abnormality classification in diagnostic images.

layer is followed by a ReLu activation layer, which is followed by a max-pooling layer excluding at the Conv3 and Conv4 layers. In between the first two ReLu and max-pooling layers, a normalization layer is added. The final three layers are adjusted accordingly to the requirements of the proposed classification model. Hence, the three layers at the rear end were removed and replaced with a fully connected, softmax, and classification output layers. The final fully connected layer is designed according to the number of classes in the new data for classification outputs. The output feature vectors are not summed; instead, the model maps the layer with the incoming features and concatenates them.

In most deep learning architectures, tanh and sigmoid are the standard activation functions normally used. However, an inherent problem with these activation function is that they have "saturating regions," and the gradient becomes very small in such regions. As a result, gradient descent becomes very slow and training takes longer, which increases the computation time. We used the rectified linear unit (ReLU) activation function, as it overcomes this gradient problem (for the positive portion). The ReLU function, given by $f(x) = \max(0,x)$, ensures that the gradient becomes zero for negative values. A special property of ReLU is that it does not require input normalization to prevent saturation. If some training examples produce a positive input to ReLU, learning is induced accordingly. However, it was found that an additional local normalization scheme aids generalization and also improves accuracy. We define the normalization scheme as per Eq. (1) and incorporate it in the proposed model. During experiments, it was observed that using ReLU with the proposed local normalization scheme improved the training by a significant amount.

$$b^i_{x,y} = a^i_{x,y} \left/ \left(k + \alpha \sum_{j=\max(0,\,i-n/2)}^{j=\min(N-1,\,i+n/2)} (a^j_{x,y})^2 \right)^\beta \right. \tag{1}$$

where $b^i_{x,y}$ represents the regularized output of kernel i at position x, y; and $a^i_{x,y}$ represents the source output of kernel i applied at position x, y. N is the total number of kernels in the layer and n is the number of the normalization neighborhood. k, α, and β are hyperparameters.

During the classification process, each class is separately fed to the input nodes of the deep neural network during training because the number of samples in each class varies as per the number of patient scan images available. For the training process, stochastic gradient descent with momentum (SGDM) was used as the solver optimizer with the learning rate initialized to 0.0001 having a batch size of 10 and the number of epochs set to 10. SGDM can switch back and forth to reach the optimum path, so the momentum parameters were added to reduce the switching problem (McHugh, 2012).

To avoid considering the same data segments at every epoch, a shuffling parameter value was set before each training epoch. That is, if the minibatch size cannot uniformly distribute the training samples, then *trainNetwork* discards the training data that do not fit into the final complete minibatch of each epoch. To learn faster in the new layers, we increased the *WeightLearnRateFactor* and *BiasLearnRateFactor* values of the fully connected layer. This was heuristically set to the rate of 20; the learning rate is determined by multiplying this factor by the global learning rate for deriving the biases in the fully connected layer. The loss function used in this CNN-based neural network is a cross-entropy loss of an encoded output. For a single image the cross-entropy loss is given by

$$Cross-entropy\ loss = \sum_{c=1}^{M} (y_c \cdot \log \widehat{y_c}) \tag{2}$$

where M is the number of classes and $\widehat{y_c}$ is the model's prediction for that class (i.e., the output of the softmax for class c). Due to the fact that the labels are encoded and y is a (2×1) vector of ones and zeros, y_c is either 1 or 0. Finally, the predicted class labels (if an input X-ray image is normal or abnormal) is obtained at the output layer.

Algorithm 1 illustrates the process of identifying the abnormal regions in musculoskeletal radiograph images. After an image is classified as normal or abnormal, an automated analysis of the type of abnormality present in the image is of critical importance. The objective is to identify potential abnormality findings such as hardware artifacts and the existence of fractures in the scan image. Boundary detection in abnormal images is one of the crucial steps while generating X-ray scan reports. Currently, the abnormal regions are marked manually by the radiographer, expert clinicians, or physicians. The developed algorithm can make a change over in the computer-aided diagnosis medical system, where the boundary will be marked by the system itself if it finds any abnormalities present in the image. In our work, a boundary detection algorithm is incorporated to detect the abnormalities present in the image. For each image, we plot the histogram of the abnormal image and binarize the image with

---

ALGORITHM 1

Abnormal region detection algorithm.

Input: An abnormal image

Output: Boundary region on abnormal image

1: **for** i = 1 to length(TestImage) **do**
2:     $I_T \leftarrow$ TestImage   ◁ *Read the test image*
3:     Compute the histogram of the image.
4:     Find total histogram values > 1000.
5:     binaryImage ← grayImage < x-axis value with last peak from histogram ◁ *Binarize the image*
6:     binaryImage ← bwareafilt(binaryImage, 2)   ◁ *Extract only the two largest blobs*
7:     labeledImage ← bwlabel(binaryImage)   ◁ *Label Connected Components*
8:     binaryImage2 = labeledImage ← 0
9:     binaryImage2 ← imfill(binaryImage2, 'holes')   ◁ *Fill holes*
10:    TestImage($\sim$ binaryImage2) ← 0   ◁ *zero out the other parts of the image*
11:    Mask ← grayImage > call   ◁ *Get a new binary image*
12:    Mask ← imfill(Mask, 'holes')   ◁ *Fill holes*
13:    Mask ← bwareafilt(Mask, 1)   ◁ *Consider largest blob*
14:    Mask ← bwconvhull(Mask)   ◁ *Take convex hull*
15:    *Get the Centroid points.*
16:    *Mark the abnormal region with the bounding box and centroid points.*
17:    Output abnormalities.
18: **end for**

---

less than the bin location, after which the largest blobs/regions in the image are determined. Next, the rightmost connected components are obtained, after which the centroid and bounding box of the masked image are captured. Using these, a bounding box is marked with the final values.

4. Experimental details and results

4.1 Dataset details

For experimental evaluation, two datasets were used—a standard open dataset called MURA (Rajpurkar, 2017) and a manually curated dataset. MURA provides musculoskeletal radiography images of seven upper extremity classes, as per the Picture Archive and Communication Systems (PACS). It consists of 14,863 studies of 12,173 patients, totaling 40,561 radiography images. The dataset is split into training (36,808 images, 13,457 studies from 11,184 patients), validation (3197 images, 1199 studies from 783 patients), and testing (556 images, 207 studies from 206 patients) sets. There is no overlap in patients between any of the sets and each study was labeled manually as either normal or abnormal by trained radiologists. Images that were labeled as abnormalities in the dataset contain anomalies such as fractures, hardware artifacts, lesions, etc. The size of the patient cohort in each of the study classes is shown in Fig. 2 and samples of normal and abnormal images from the dataset are given in Fig. 3. The manually curated dataset consists of CXR images collected from local hospitals and also from publicly available open access sources for the classification and detection of COVID-19 from frontal chest radiographs. It consists of two classes, that is, COVID-19 and normal/no findings—each consisting of 120 frontal CXR images of different patients. Table 1A and B shows statistics relating to the CXR and MURA dataset. Table 1 summarizes the training and validation/testing set distribution for normal and abnormal studies.

The proposed model was evaluated for each of the study classes and also on the overall classes for abnormality classification. First, the predicted class labels were obtained from a

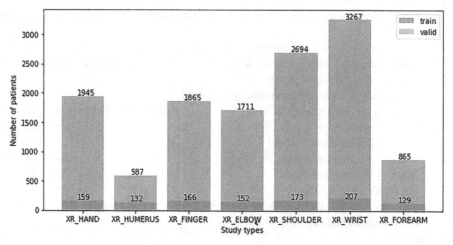

FIG. 2   Patient counts per study type.

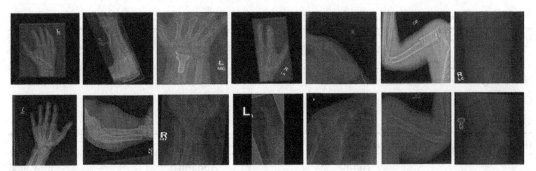

FIG. 3 Sample images of the MURA dataset in hand, forearm, wrist, finger, shoulder, humerus, and elbow classes (*upper row*—abnormal images; *lower row*—normal images).

TABLE 1 Dataset: Statistics.

(A) Curated CXR dataset: statistics

Class	Training		Testing		Total
			Normal	Abnormal	
Chest	Cross	validation	120	120	240

(B) MURA dataset: statistics

Classes	Training		Testing		Total
	Normal	Abnormal	Normal	Abnormal	
Elbow	1094	660	92	66	1912
Finger	1280	655	92	83	2110
Forearm	590	287	69	64	1010
Hand	1497	521	101	66	2185
Humerus	321	271	68	67	727
Shoulder	1364	1457	99	95	3015
Wrist	2134	1326	140	97	3697
Total	8280	5177	661	538	14,656

deep CNN model. The confusion matrices for actual and predicted labels are shown in Table 2. For CXR images, a fivefold cross validation was performed using the DenseNet model. Out of 240 CXR images, both classes consisted of 120 images each. Among those, 103 images were classified as no findings and 109 images were classified as COVID-19. In the MURA dataset, a total of 1199 upper extremity musculoskeletal images were classified, out of which 661 images were normal and 538 images were abnormal findings (including fractures, hardware artifacts, and joint diseases). Of these, 79.18% of images were correctly classified as normal and 66.67% of images were predicted to be abnormal.

TABLE 2 Confusion matrix.

(A) Confusion matrix on curated CXR

	No findings	COVID-19	Row total
No findings	103	17	120
COVID-19	11	109	120
Col. total	114	126	240

(B) Confusion matrix on MURA dataset

	Normal	Abnormal	Row total
Normal	524	137	661
Abnormal	179	359	538
Col. total	703	496	1199

The performance of the proposed model was measured using standard evaluation metrics such as accuracy, sensitivity, specificity, and kappa statistics. The accuracy of a test can be defined as its ability to differentiate the patient's normal and abnormal cases correctly. The accuracy of a test is estimated by calculating the proportion of true positive and true negatives in all evaluated cases, as per Eq. (4). Sensitivity and specificity are the statistical measures of the performance for a binary classification task, calculated, as per Eqs. (5), (6). Sensitivity measures the proportion of actual positives that are correctly identified as positive (i.e., the percentage of abnormalities in the study that are correctly identified as abnormalities). Specificity measures the proportion of actual negatives that are correctly identified as negative (i.e., the percentage of the normal study that is correctly identified as normal).

$$Accuracy = \frac{TP + TN}{TP + TN + FP + FN} \tag{3}$$

$$Sensitivity\ (TPR) = \frac{TP}{TP + FN} \tag{4}$$

$$Specificity\ (TNR) = \frac{TN}{TN + FP} \tag{5}$$

Cohen's kappa statistics (McHugh, 2012) measure the interobserver agreement or precision, and are especially useful when the same score is assigned to the same data items. This statistic is calculated when the data scores two trials on each sample, that is, a binary classification. Because the dataset used in this model is a binary classification model, the outcome is to predict whether the data sample under test is normal or abnormal. The importance of this metric lies in the correct representation of the data measured. Its range is −1 to +1; a value of 1 indicates a perfect agreement and values less than 1 indicate less than a perfect agreement. In some rare situations, kappa value can also be negative, which is a sign that the agreement score is much lower than the expected rate. It is computed as per Eq. (6), where $Pr(a)$ is the actual observed agreement, and $Pr(c)$ is the chance agreement.

TABLE 3 Observed classification performance with respect to different classes.

(A) Curated CXR dataset results

Classes	Accuracy	Sensitivity	Specificity	F1 score
Chest	89.58	85.83	90.83	88.33

(B) MURA dataset results

Classes	Accuracy	Sensitivity	Specificity	Kappa
Elbow	76.99	83.83	70.00	0.539
Finger	69.85	78.97	61.94	0.403
Forearm	72.09	85.33	58.94	0.442
Hand	70.65	84.13	51.32	0.369
Humerus	78.47	79.05	77.86	0.569
Shoulder	68.56	68.77	68.35	0.371
Wrist	79.67	88.74	68.47	0.581
Overall	**73.19**	**79.18**	**66.67**	**0.46**

$$kappa = \frac{Pr(a) - Pr(c)}{1 - Pr(c)} \tag{6}$$

The performance obtained for each of the classes is tabulated in Table 3. It was found that the model performed its best in classifying the CXR images with an accuracy of 89.58%, and sensitivity and specificity scores of about 85.83% and 90.83%, respectively. In another case, it was seen that the model achieved good results for the *elbow, forearm, hand, humerus,* and *wrist* classes, achieving >70% accuracy. However, the *finger* and *shoulder* classes showed lower accuracy. The other two metrics used, sensitivity and specificity, capture some additional aspects of the classification performance. Sensitivity is a measure of the true positive rate, or probability of detection. In this case, it indicates the percentage of medical scans correctly identified as abnormal. Specificity or the true negative rate gives the percentage of normal medical scans that was correctly classified as normal. From Table 3B, the observed sensitivity scores for the *elbow, forearm, hand,* and *wrist* classes were in the range of 84%–89%, indicating that for these classes the abnormal samples were correctly classified to a larger extent. However, the specificity scores of classes such as *forearm, hand,* and *finger* indicate that the percentage of normal scans correctly classified as normal was lower than that of the other classes. The lowest sensitivity and specificity scores were observed for the *shoulder* class, indicating that the model was not able to distinguish between normal and abnormal images in this class very well, thus requiring more detailed scrutiny and analysis. On average, the proposed approach achieved an accuracy rate of 73.19%, with sensitivity and specificity scores of 79.18% and 66.67%, respectively, which indicates good classification performance. A graphical plot that illustrates the diagnostic ability of a classifier system in terms of the area under the receiver operator characteristic (ROC) curve (AUROC) is illustrated in Fig. 4. An AUROC value of 95.84 is achieved, indicating good

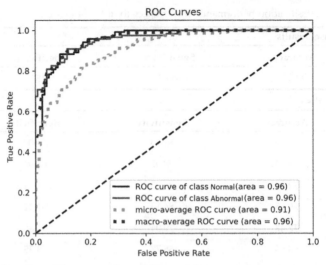

FIG. 4 AUROC performance of the proposed ensemble model.

performance in distinguishing anomalous and nonanomalous radiographical scans. As discussed earlier, the average kappa statistic value was 0.46, which indicates moderate agreement on the test samples with the expected values. However, radiograph readings and their findings are often judged subjectively, so other metrics such as accuracy, specificity, and sensitivity are also used to gain more refined insights into the performance of the proposed model.

The proposed model was benchmarked against DenseNet, MobileNet, NASNETmobile, and Xception models employed by other researchers, the results of which are summarized in Table 4. Rajpurkar (2017) used only one evaluation metric, *kappa*, making it difficult to benchmark our model with theirs with reference to other standard evaluation metrics such as accuracy, sensitivity, and specificity. The datasets used in our work provide X-ray images, which do not contain any explicit region-of-interest annotations, so we used Grad-CAM visualizations to demarcate regions of interest for CXR images when the abnormality detection method is applied for upper extremity musculoskeletal images. The model visualization is overlaid as a heatmap for identifying the predicted abnormalities in CXR, and this is shown in Fig. 5. The results of the proposed abnormality detection algorithm are depicted in Fig. 6 for

TABLE 4 Benchmarking the proposed model against state-of-the-art works.

Models	Layers	Accuracy	Sensitivity	Specificity	Kappa
Proposed model	**25**	**0.73**	**0.79**	**0.66**	**0.46**
EnsembleE *[Xcep, Dense] (Banga and Waiganjo, 2019)*	>169	0.71	0.77	0.63	0.41
MobileNet *(Single)* (Banga and Waiganjo, 2019)	88	0.67	0.73	0.61	0.34
EnsembleD *[Xcep, MobileN]* (Banga and Waiganjo, 2019)	>169	0.65	0.73	0.56	0.29
EnsembleA *[NASN, Xcep,MobileN]* (Banga and Waiganjo, 2019)	>169	0.57	0.67	0.46	0.13

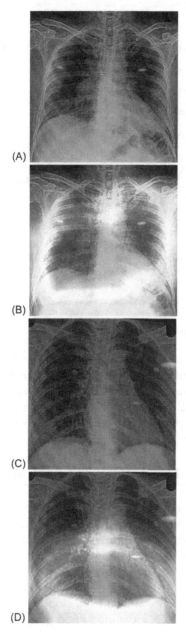

FIG. 5 Model's grad-CAM (heat-map) visualizations for COVID-19. (A) COVID-19 image; (B) grad-CAM; (C) COVID-19 image; and (D) grad-CAM.

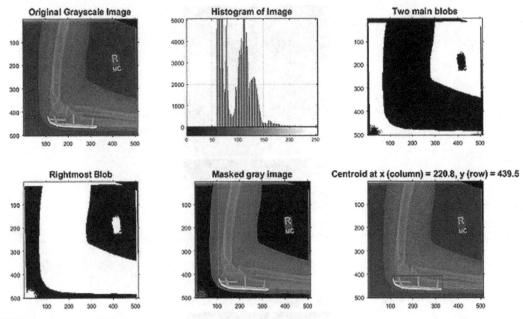

FIG. 6   Detection of abnormality regions in elbow class.

the elbow class of musculoskeletal images. Hardware artifacts used for setting bones such as metal inserts and screws are automatically detected and correctly classified as a type of abnormality. Similarly, even fractures and cracks were also captured correctly as abnormalities using our abnormal detection algorithm. During experimental validation, we found that the bounding detection algorithm performed well, which is evident from Fig. 6.

5. Conclusion and future work

In this chapter, a deep neural model for classifying medical scan images into normal or abnormal as well as identifying the type of abnormalities present in an image were presented. We developed an abnormal region detection algorithm and used it for identifying the anomalous regions in the image. As the scan images vary in resolution and the type of studies available in each class is different, the proposed deep CNN model was trained separately for each of the classes. Experimental performance evaluation revealed that the proposed model achieved accuracies of 89.58% and 73.19%, along with promising sensitivity, specificity, and kappa statistic value of 0.46, indicating good performance. The overall computation time was also reduced during training, as our proposed model uses a comparatively shallow architecture when compared to state-of-the-art models. These models used a much deeper architecture (shown in Table 4). However, the proposed model still outperformed them while being less computationally expensive due to lesser training time.

During experiments, it was observed that the proposed approach under performed in some unbalanced classes such as shoulder, finger, and hand, which needs to be further analyzed and improved upon for enhancing the proposed model. Also, we intend to explore different image enhancement techniques for addressing additional faults such as blur, tilt, over/under exposure, etc. Integrating contrast adjustment and noise resolution by super resolving the scan images for improving the detection of abnormal regions will also be explored.

References

Alhudhaif, A., Polat, K., Karaman, O., 2021. Determination of COVID-19 pneumonia based on generalized convolutional neural network model from chest X-ray images. Expert Syst. Appl. 180, 115141.

Banga, D., Waiganjo, P., 2019. Abnormality detection in musculoskeletal radiographs with convolutional neural networks (ensembles) and performance optimization. ArXiv preprint arXiv:1908.02170.

BMUS, 2014. United States Bone and Joint Initiative: The Burden of Musculoskeletal Diseases in the United States (BMUS), fourth ed. http://www.boneandjointburden.org/2014-report.

Ching, T., Himmelstein, D.S., Beaulieu-Jones, B.K., Kalinin, A.A., et al., 2018. Opportunities and obstacles for deep learning in biology and medicine. J. R. Soc. Interface 15 (141), 20170387.

Fang, Y., Zhang, H., Xie, J., Lin, M., Ying, L., Pang, P., Ji, W., 2020. Sensitivity of chest CT for COVID-19: comparison to RT-PCR. Radiology 296 (2), E115–E117.

Gale, W., Oakden-Rayner, L., Carneiro, G., Bradley, A.P., Palmer, L.J., 2017. Detecting hip fractures with radiologist-level performance using deep neural networks. ArXiv preprint arXiv:1711.06504.

Hajabdollahi, M., Esfandiarpoor, R., Sabeti, E., Karimi, N., Najarian, K., Soroushmehr, S., Samavi, S., 2018. Multiple abnormality detection for automatic medical image diagnosis using bifurcated convolutional neural network. ArXiv preprint arXiv:1809.05831.

Huang, G., Liu, Z., Van Der Maaten, L., Weinberger, K.Q., 2017. Densely connected convolutional networks. In: Proceedings of the IEEE Conference on Computer Vision and Pattern Recognition, pp. 4700–4708.

Jain, R., Gupta, M., Taneja, S., Hemanth, D., 2021. Deep learning based detection and analysis of COVID-19 on chest X-ray images. Appl. Intell. 51 (3), 1690–1700.

Karargyris, A., Siegelman, J., Tzortzis, D., Jaeger, S., et al., 2016. Combination of texture and shape features to detect pulmonary abnormalities in digital chest X-rays. Int. J. Comput. Assist. Radiol. Surg. 11 (1), 99–106.

Karthik, K., Kamath, S., 2021. Deep neural models for automated multi-task diagnostic scan management—quality enhancement, view classification and report generation. Biomed. Phys. Eng. Express 8 (1), 015011.

Karthik, K., Kamath, S., 2022. MSDNet: a deep neural ensemble model for abnormality detection and classification of plain radiographs. J. Ambient Intell. Humaniz. Comput. 13 (5), 1–14.

Krizhevsky, A., Sutskever, I., Hinton, G.E., 2012. Imagenet classification with deep convolutional neural networks. Adv. Neural Inform. Process. Syst. 25, 1097–1105.

Lai, Z., Deng, H., 2018. Medical image classification based on deep features extracted by deep model and statistic feature fusion with multilayer perceptron. Comput. Intell. Neurosci. 2018, 2061516.

Lytras, M, Sarirete, A, Visvizi, A., Chui, KT., 2021a. Artificial Intelligence and Big Data Analytics for Smart Healthcare. Academic Press, eBook, ISBN: 9780128220627. https://doi.org/10.1016/C2019-0-03287-6.

Lytras, D.M., Lytra, H., Lytras, M.D., 2021b. Chapter 1—Healthcare in the times of artificial intelligence: setting a value-based context. In: Lytras, M.D., Sarirete, A., Visvizi, A., Chui, K.T. (Eds.), In Next Gen Tech Driven Personalized Med&Smart Healthcare, Artificial Intelligence and Big Data Analytics for Smart Healthcare. Academic Press, ISBN: 9780128220603, pp. 1–9. 2021 https://doi.org/10.1016/B978-0-12-822060-3.00011-5.

Mayya, V., Karthik, K., Kamath, S.S., Karadka, K., Jeganathan, J., 2021. COVIDDX: AI-based clinical decision support system for learning COVID-19 disease representations from multimodal patient data. In: HEALTHINF.

McHugh, M.L., 2012. Interrater reliability: the kappa statistic. Biochem. Med. 22 (3), 276–282.

Narin, A., Kaya, C., Pamuk, Z., 2021. Automatic detection of coronavirus disease (COVID-19) using X-ray images and deep convolutional neural networks. Pattern Anal. Appl. 24 (2), 1–14.

Osman, A.H., Aljahdali, H.M., Altarrazi, S.M., Ahmed, A., 2021. SOM-LWL method for identification of COVID-19 on chest X-rays. PLoS One 16 (2), e0247176.

Rahmany, I., Nemmala, M.E.A., Khlifa, N., Megdiche, H., 2019. Automatic detection of intracranial aneurysm using LBP and Fourier descriptor in angiographic images. Int. J. Comput. Assist. Radiol. Surg. 14 (8), 1353–1364.

Rajpurkar, P., 2017. MURA dataset: towards radiologist-level abnormality detection in musculoskeletal radiographs. ArXiv preprint arXiv:1712.06957.

Rajpurkar, P., Irvin, J., Zhu, K., Yang, B., Mehta, H., Ng, A., 2017. Chexnet: radiologist-level pneumonia detection on chest X-rays with deep learning. ArXiv preprint arXiv:1711.05225.

Sait, U., Kv, G.L., Shivakumar, S., Kumar, T., Bhaumik, R., Prajapati, S., Bhalla, K., Chakrapani, A., 2021. A deep-learning based multimodal system for COVID-19 diagnosis using breathing sounds and chest X-ray images. Appl. Soft Comput. 109, 107522.

Selvaraju, R., Cogswell, M., et al., 2017. Grad-cam: visual explanations from deep networks via gradient-based localization. In: IEEE International Conference on Computer Vision, pp. 618–626.

Shakarami, A., Menhaj, M., Tarrah, H., 2021. Diagnosing COVID-19 disease using an efficient CAD system. Optik 241, 167199. https://doi.org/10.1016/j.ijleo.2021.167199.

Shen, D., Wu, G., Suk, H.-I., 2017. Deep learning in medical image analysis. Annu. Rev. Biomed. Eng. 19, 221–248.

Silvian, S.P., Maiya, A., Resmi, A., Page, T., 2011. Antecedents of work related musculoskeletal disorders in software professionals. Int. J. Enterprise Netw. Manag. 4 (3), 247–260.

Spruit M., Lytras M., (2018). Applied data science in patient-centric healthcare: adaptive analytic systems for empowering physicians and patients, Telemat. Inform., 35(4), 2018, 643–653, https://doi.org/10.1016/j.tele.2018.04.002.

Tataru, C., Yi, D., Shenoyas, A., Ma, A., 2017. Deep learning for abnormality detection in chest X-ray images. In: IEEE Conference on Deep Learning.

Wærsted, M., Hanvold, T.N., Veiersted, K.B., 2010. Computer work and musculoskeletal disorders of the neck and upper extremity: a systematic review. BMC Musculoskelet. Disord. 11 (1), 79.

Wang, L., Lin, Z.Q., Wong, A., 2020. Covid-net: a tailored deep convolutional neural network design for detection of COVID-19 cases from chest X-ray images. Sci. Rep. 10 (1), 1–12.

WHO, 2020. Coronavirus disease (COVID-19). https://www.who.int/health-topics/coronavirus#tab=tab_3.

Xi, P., Shu, C., Goubran, R., 2018. Abnormality detection in mammography using deep convolutional neural networks. In: 2018 IEEE International Symposium on Medical Measurements and Applications (MeMeA), IEEE, pp. 1–6.

Smart healthcare digital transformation during the Covid-19 pandemic

Roberto Moro-Visconti

Department of Economics and Business Management Sciences, Università Cattolica del Sacro Cuore, Milan, Italy

1. Introduction

The COVID-19 pandemic has challenged healthcare systems worldwide. Uncertainty of transmission, limitations of physical healthcare system infrastructure and supplies, and workforce shortages require the dynamic adaption of resource deployment to manage rapidly evolving care demands, ideally based on real-time data for the entire population. Moreover, shutting down the traditional face-to-face care infrastructure requires the rapid deployment of virtual healthcare options to avoid the collapse of health organizations (Baumgart, 2020).

Healthcare is responding to the COVID-19 pandemic through the fast adoption of digital solutions and advanced technology tools that make it "smart" (as previously described by Gastaldi and Corso, 2012).

The COVID-19 pandemic, like all serious disruptions in human history, is causing unprecedented health and economic crises. At the same time, though, this new situation is favoring digital transition in many industries and societies. This is the case, for example, with education. The entire sector, from primary schools to universities, has developed new strategies for teaching remotely, shifting from lectures in classrooms to live conferencing or online courses. Similarly, healthcare is responding to the COVID-19 pandemic through the fast adoption of digital solutions and advanced technology tools. In times of pandemic, digital technology can mitigate or even solve many challenges, thus improving healthcare delivery. This is currently being done to address acute needs that are a direct or indirect consequence of the pandemic (e.g., apps for patient tracking, remote triage emergency services, etc.). Nevertheless, many of the solutions that have been developed and implemented during the emergency could be consolidated soon, contributing to the definition and adoption of new digital-based models of care (Golinelli et al., 2020).

Within this evolving framework, digital technologies are thriving, and their scalable impact is catalyzed by concomitant events, affecting complex and chaotic healthcare systems.

There is no doubt that COVID-19 has been a terrible scourge, but it has also stamped down on the accelerator for tech areas such as home broadband, collaboration services, cloud applications, and services generally, along with remote working. But what COVID did to digital healthcare may have the most consequential of impacts (https://www.itu.int/en/myitu/News/2021/04/07/07/25/COVID-accelerating-digital-healthcare).

According to Herrmann et al. (2018), digital technologies can be categorized according to the healthcare needs they address: diagnosis, prevention, treatment, adherence, lifestyle, and patient engagement.

Medical 4.0 (personalized therapy; artificial intelligence-enabled devices; Internet of Medical Things, etc.), if implemented, can adequately handle the ongoing situation in the medical field as it will provide applications of advanced technologies to take care of the challenges of the COVID-19 outbreak (Haleem and Javaid, 2020).

Digital health information technologies can be applied in three aspects, namely digital patients, digital devices, and digital clinics, and could be useful in fighting the COVID-19 pandemic (Kalhori et al., 2021).

The COVID-19 pandemic has radically and quickly altered how medical practitioners provide care to patients. Medical centers are now responding to COVID-19 through the rapid adoption of digital tools and technologies such as telemedicine and virtual care, which refer to the delivery of healthcare services digitally or at a distance using ICT for the treatment of patients. Telemedicine is expected to deliver timely care while minimizing exposure to protect medical practitioners and patients (Bokolo Jr., 2020).

Even if healthcare is traditionally strongly dependent on technological upgrades, digital applications are still in their infancy, although they have enormous potential. Pervasive digitization of the supply and value healthcare chains represents a disruptive innovation able to reshape the whole healthcare ecosystem, reducing the distance among patients, health centers, and physicians.

The literature streams are wide and interdependent, as they connect different subjects innovatively, as shown in Fig. 1.

Although the literature in all four fields is enormous, some references may be summarized as follows:

(a) Network theory; see Barabási, 2016; Bianconi, 2018.
(b) Digitalization (Asadullah et al., 2018; Moro Visconti, 2020; Sutherland and Jarrahi, 2018).
(c) M-Health/e-Health (Dossi et al., 2020); applications to Covid pandemics (Nishi et al., 2020; Swanson and Santamaria, 2021; Davis, 2020; Kalhori et al., 2021).
(d) Digital transformation in healthcare (Kraus et al., 2021; Deiters et al., 2018; EXPH (Expert Panel on Effective Ways of Investing in Health), 2019; Gastaldi and Corso, 2012; Herrmann et al., 2018; Iyawa et al., 2017).

The chapter is structured as follows: Section 2 contains an innovative description of the interaction between viral networks and digital platforms, using graph (network) theory, after

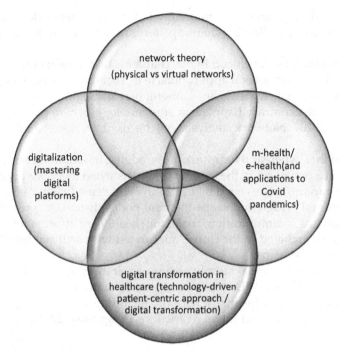

FIG. 1   Interacting literature streams.

having recalled the main epidemics in history; Section 3 illustrates the workings of digital platforms and their economic rationale before a network theory interpretation of digital nodes in Section 4. Section 5 illustrates the impact of scalable intangibles (big data, artificial intelligence, blockchains, etc.) on healthcare. Section 6 contains a discussion and Section 7 some concluding remarks.

2. Viral networks and digital platforms during the Covid-19 pandemic

COVID-19, belonging to the coronavirus family, has unleashed a global pandemic ($\pi\alpha\nu+\delta\varepsilon\mu o\sigma$=all the people) that, due to its magnitude and effects, brings back memories of ancestral scourges, reproposing ancient threats that we considered confined to the history books. The link between man and nature has always been conflictual and punctuated by biblical pestilences that periodically resurface, subverting an order that *Sapiens* (see Harari, 2015) claims in vain to dominate.

The black swan reappears unexpectedly, symbolizing an unforeseen event that distorts the trajectory of our expectations, and which, in retrospect, is improperly rationalized and judged predictable.

And the pandemic spreads through a network made up of physical contiguity and social relationships, catalyzed by globalization and connected to an increasingly pervasive digital dimension.

In this context, still in progress and hopefully fading, viral networks intersect with platforms and virtual networks, following chaotic patterns that deserve to be interpreted and then mastered.

A key concept that will be developed in this chapter is that networks represent a powerful way of communication, with several applications (e.g., social networks, energy transmission, transportation, food chains). Whereas links among networking nodes typically have a positive economic value, the contrary happens in epidemiology, where pathogen links transmit virality (So et al., 2020). Network analysis can be used to assess global pandemic risk (So et al., 2021).

It will be shown that, whereas the rupture of the edging links among nodes causes a problem (for instance, in a power grid), the contrary happens when there is a pandemic outbreak. E-health (with the variants represented by m-health and telemedicine) can soften these problems, building a virtual bridge that keeps the viral nodes distant, so preventing contagion. Digitalization maintains communication, coalescing healthcare information that can be gathered to form big data, archived in the cloud, interpreted with artificial intelligence patterns, and validated with blockchains.

Pandemics impact international supply chains, generating several criticalities (Swanson and Santamaria, 2021). The COVID-19 pandemic has brought attention to supply chain networks because of disruptions for many reasons, including labor shortages because of illnesses, death, risk mitigation, and travel restrictions (Nagurney, 2021).

2.1 Epidemic plagues in literature and history

An understanding of the value and the dangers attributable to epidemic networks can be acquired through historical study.

There have been three great world pandemics of plague recorded, in 541 CE, 1347–1348 CE (the Black Death), and 1894 CE, each time causing devastating mortality of people and animals across nations and continents. On more than one occasion, plague irrevocably changed the social and economic fabric of society (Frith, 2012).

The plague plays a leading role in the collective imagination of epidemics, which has been handed down to us by great writers from different eras.

Homer, in the Iliad, speaks to us of a "feral disease," a divine punishment against the Greeks triggered by the behavior of Agamemnon, guilty of having offended the priest Chryses. The book of Numbers tells us of a pestilence in the time of Moses, also of divine origin. Thucydides in the Peloponnesian War describes the Athens plague of 430 BCE, which also killed Pericles. Lucretius takes up the episode in the sixth book of *De Rerum Natura*. Tacitus in his Annales describes the Rome epidemic of 66 CE. Procopius of Caesarea in his Secret History tells us about the plague of the Byzantine emperor Justinian who always struck Rome.

Giovanni Boccaccio in the *Decameron* makes the story of the black plague of 1348 lighter, even if it killed at least a third of the European population. It had almost 20 million victims, plunging the continent into an epochal depression that lasted until the Renaissance, becoming one of the most compelling events in human history (Kelly, 2005).

The historian and economist Carlo Maria Cipolla (1981) describes with detached irony the plague and health facilities in Tuscany.

Daniel Defoe in the *Journal of the Plague Year* recalls the disease that struck London in 1665 that he witnessed as a child.

Alessandro Manzoni describes in the *"Promessi sposi"* the epic of the plague in Milan, Italy, in the 1600s, which entered with the German bands (as it seems COVID-19 spread to Lombardy in January 2020, through patient 1 infected by a German who went to China) and then invaded and depopulated a large part of Italy. Albert Camus tells us about the Algerian plague of the 1940s.

The plague has been followed by the "Spanish" influenza, which spread a flu virus that between 1918 and 1920 killed tens of millions of people around the world, infecting about 500 million people out of a world population of about 2 billion. This was a slow and distracted contagion from the aftermath of the great war, which spread with the means of transport—and was told with the media—of that time.

In more recent times, there have been other viral epidemics that have not affected our borders, acting as unheard alarm bells. These include MERS, triggered by another coronavirus starting in April 2002, and SARS, a viral pneumonia that first appeared in November 2002 in Guangdong province in China. This also includes the 2009–2010 Influenza A pandemic (swine fever), which has also affected Italy. Nor should we forget the viral hemorrhagic fevers endemic in sub-Saharan Africa, such as Ebola (which takes its name from the Congo River where it was discovered), which plagued West Africa in 2014–2016. It also reappears periodically, claiming victims among the poorest.

2.2 Könisberg bridges and graph theory

The theory of graphs or networks, consistent with the theoretical analysis of e-health patterns, has been used for some decades to develop mathematical models that interpret, among other things, the spread of viruses, trying to predict their developments (Barabási, 2016). Before summarizing its features, it is worthwhile to recall the story of discovery, intuitive , but important and versatile, not only in the epidemiological field.

The problem concerns the possibility of making a route crossing all seven bridges of the city of Kant (today Kaliningrad), passing each bridge only once. The question (with a negative answer, not very relevant but the harbinger of a great intuition) was solved by the Swiss mathematician Euler with the *Solutio problematis ad geometriam situs pertinentis* of 1741; in it today we identify the origin of the modern theory of graphs. Graphs are vertices or nodes connected through sides, just like bridges to cross.

The structure of the nodes forms a network.

The applications of network theory are vast and interdisciplinary, ranging from statistical or particle physics to computer science (the Internet is a large virtual network), electrical engineering (electrical transmission networks), biology (food chains), economics, finance (financial market networks, operators, and intermediaries), operational research, climatology (interrelationships between clouds, wind, temperatures, and other factors), and sociology (social networks, etc.).

Networks have an invasive presence in everyday life. For example, think of the interconnections related to the exchange of e-mails or messages, the use of social networks, telephone

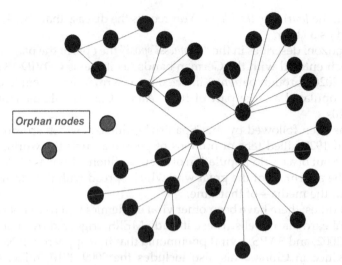

FIG. 2 Example of a network.

calls, transport, money transfers, epidemics, food chains, ecosystems, electricity grids, etc. In all these cases, networks and their properties are used.

Disorderly or random modes of interaction connect individuals or things, with relevant consequences from a legal point of view. Some recent network applications with no hierarchical structure concern blockchains.

A network represents, in its most elementary formulation, a set of points (vertices or nodes) connected by lines (sides or edges).

Fig. 2 illustrates a network with two types of nodes, central or peripheral. The central nodes have a greater number of connections (not only with the other nodes but as a bridge between one node and another) and are consequently the most valuable.

The red nodes in the middle left of the graph are "orphans." They are good news in epidemics (as they exemplify a person detached from others, thanks to social distance or immunity), but not so in many other networks (social networks; e-commerce applications, but also telemedicine or e-health) where value depends on connectivity. Interconnectivity naturally implies vulnerability. If nodes are disconnected, then a path with other vertices does not exist.

2.3 Viral networks

Among the various applications, some models have been used to study viral networks and their epidemiological features for several decades. The basic assumption is simple, configuring, for example, two nodes (human beings) connected by a virus. Concepts such as the now known R_0 contagion coefficient are born from this.

The basic reproduction number, R_0, indicates in epidemiology the potential transmissibility of infectious disease, expressing the number of new cases generated on average by a single case during its infectious period. The parameter varies considerably in the different pathologies (it is very high, for example, in measles) and expresses the "virality" of each node (infected patient).

The applications of network theory to virology, known for decades, are extremely relevant and can allow a forecast mapping of infections, suggesting tools and methods of containment.

Epidemics can spread across large regions to become pandemics by flowing along with transportation and social networks. Two network attributes, transitivity (when a node is connected to two other nodes that are also directly connected between them) and centrality (the number and intensity of connections with the other nodes in the network), are widely associated with the dynamics of pathogen transmission (Gómez and Verdú, 2017).

In general terms, it can be noted that each node has mathematical properties that measure its importance. For example, in transport, the Bologna railway node is the most important in Italy because it acts as an interchange and passage point for heavily trafficked routes.

Generally, the insertion of an additional node adds value to the network. This surplus value is measurable with the model of Metcalfe (the inventor of the Ethernet network), according to which the increase in value is exponential.

The issue can be understood, in opposite terms, by assuming the loss of value deriving from the absence of a node. This can explain, for example, the electrical blackouts that occur when a node in the transmission chain breaks. At 3:27 a.m. on Sept. 28, 2003, the whole of Italy ran out of electricity except the islands of Capri and Sardinia, which had autonomous networks. The problem was caused by the fall of a tree in Switzerland, which broke the network between France and Italy. Blackouts are a typical example of cascading failure with domino effects.

In most cases, the disappearance or cancellation of a node creates potentially significant problems, especially in physical networks, triggering domino effects and chain reactions that are not always predictable.

In the epidemiological field, considering for example viral networks, exactly the opposite happens. Isolation between nodes (individuals), implemented through social distancing, protective masks, constant disinfection, etc., are all measures that are part of a strategy aimed at interrupting or at least loosening the viral contagion among multiple nodes. They are physical barriers against an invisible enemy, much smaller than bacteria, identifiable only with electron microscopes. Most of the viruses that have been studied have a diameter between 20–300 nm.

The virus ("poison" in Latin) is a biological entity with the characteristics of an obligate parasite, which replicates exclusively inside the cells of other host organisms.

Viral networks express the mechanisms of multiplication of these pathogens through networks fed by the contiguity of individuals and by life models resulting from globalization and mass socialization.

Comparative analyses on the spreading times of Spanish flu a little more than a century ago (a time of about a year and a half) and COVID-19 of 2020 make us understand the dimensions of the problem and the acceleration suffered in a century because of the demographic multiplier (today's population is more than 7.8 billion inhabitants, as the *worldometers* statistical site reports in real time).

Information about a pandemic outbreak can nowadays be easily distorted by fake news. Infodemic risk indices estimate the chance that a user in a social media platform is pointed to potentially unreliable sources of misinformation or disinformation about COVID-19 (https://covid19obs.fbk.eu/#/).

As Davahli et al. (2021) showed, the COVID-19 pandemic has had unprecedented social and economic consequences in the United States. Therefore, accurately predicting the dynamics of the pandemic can be very beneficial. Two main elements required for developing reliable predictions include: (1) a predictive model, and (2) an indicator of the current condition and status of the pandemic. As a pandemic indicator, we used the effective reproduction number (Rt), which is defined as the number of new infections transmitted by a single contagious individual in a population that may no longer be fully susceptible. To bring the pandemic under control, Rt must be less than one. To eliminate the pandemic, Rt should be close to zero. Therefore, this value may serve as a strong indicator of the current status of the pandemic. For a predictive model, we used graph neural networks, a method that combines graphical analysis with the structure of neural networks.

2.4 The myth of the cave

The often subliminal link between the physical world and virtual reality has a profound and tendentially increasingly pervasive impact on the psyche and human behavior. The ephemeral historicity of interrelations that still smell of fresh paint does not yet allow for reasoned analysis. However, some preliminary interpretations can be attempted, to provoke more thoughtful discussion.

The myth of the cave represents one of Plato's best-known metaphors, illustrated in Chapter VII of the dialogue of the *Republic*. Its profound meaning may suggest an interesting interpretative key of the interrelation between empirical reality and the virtual.

An imaginary interview that the philosopher Giovanni Reale did with Plato several years ago comes to mind. Reale (1990) metaphorically asked Socrates' pupil if he did not find that the chained men who saw nothing but the images in the background of the cave and listened to the echo of voices coming from outside could dramatically represent the situation in which they find themselves. Men of today only glimpse images transmitted by televisions, computers, and the Internet by various multimedia communication tools, mostly from a virtual point of view, which abstracts from reality to the point of deforming its original archetypes.

In the perception of reality, there is a hierarchy between an external world (very different from that imagined by the freed Prometheus who manages to get out of the cave), and the world reflected inside the cave.

At first, reinterpreting the myth, it would seem clear that virtual reality is inside the cave and the "real" physical world outside it. But today this antinomy appears more nuanced and leads to questioning a physical reality that is confused with the virtual one to the point of being dominated by it. This is associated with the increasingly pervasive presence of iconic avatars who project their (retouched) physical image into an increasingly abstract digital environment, which can be manipulated with artificial intelligence algorithms whose applications and purposes we struggle to recognize.

3. Digital platforms

Alongside viral networks and well-known physical networks (from the food chain to transport networks, etc.), there are others, more recently introduced. We refer first to digital networks, in which physical nodes (consumers, businesses, etc.) are connected through bridge nodes such as digital platforms.

The applications are the most varied and range from e-commerce sites to domain names of websites or to mobile apps that represent the iconic Internet access portal. The value of a node is a function of the Internet traffic flows that it can carry and mediate. The traffic of information (small data aggregated to form big data) or commercial transactions, detected in real time and with the gift of ubiquity (anytime, anywhere, "24/7").

Digital platforms (Asadullah et al., 2018; Sutherland and Jarrahi, 2018) are typical nonrival goods (a sandwich is a rival good, which can only be consumed by one person, while the platform can also be used simultaneously by an unlimited number of users). And as nonrival goods, platforms have a scalable value that grows exponentially and benefits from economies of experience, incorporating the feedback of consumers who participate in a process of cocreation of value (think of the TripAdvisor site).

Smart working or e-learning (which most university teachers and students have recently discovered, familiarizing themselves with digital platforms for distance learning) are further examples of ecosustainable applications that are based on digital platforms. They are unfit to supplant frontal teaching, but able to represent, in a context of lockdown, an irreplaceable substitute.

At the base of digital platforms, there are mechanisms of interchange between node users based on the theory of networks. And the platforms add value to the network, acting as a bridge node (like an airport hub that concentrates and then routes traffic to peripheral airport nodes).

There are significant similarities and differences between viral networks and digital platforms. First, in both cases, it is a question of interpretable interrelationships with network theory.

The differences are significant. Viral networks have a physicality, albeit invisible, that is lacking in the intangible digital reality. The nodes have an opposite value: in the epidemiological field, an attempt is made to destroy their interrelation, which instead takes on a surplus value from a digital perspective. Physical distance—social distancing—is a positive element in the epidemiological field that we try to overcome and substitute with digital social networking.

Alongside these antinomies, there are also factors of contiguity that occur when the real world, in its physical and even epidemiological nature, intersects, in a not always orderly way, with the virtual one.

The interrelationships between physical (sometimes viral) and virtual networks can also be mathematically interpreted through innovative theories of the so-called multilayer network, in which networks positioned on different planes (physical and/or virtual) can intersect through "bridge nodes" present in several layers. And the kinetic value of networks must also be evaluated.

In practical terms, the interrelationships are also based on functional and organizational characteristics, based on which virtuous osmosis can be implemented that aims at loosening viral networks by strengthening virtual ones in parallel. Thus, for example, the enhancement of smart working, in its multiple articulations, to consolidate social distancing through a virtual "plexiglass." Contact tracing, with the geolocation of COVID-positives, can usefully complement other containment measures and is based on geographic information systems that map dynamic networks. Digital platform coordination is a prerequisite for any successful application.

Moreover, the digital divide does not allow a truly global application of digital services, affecting vast areas that are technologically backward and often overpopulated and that can only make limited use of applications such as teleworking, big data, artificial intelligence, data validation via blockchain, e-commerce based on B2B2C digital platforms, etc. Also, for this reason (and for the shortcomings of health systems), pandemics in the most backward areas of the world are claiming many more victims, as Ebola has shown. And epidemics involving large slums (such as Kibera in Nairobi) incorporate very precarious hygienic conditions, outclassed by other existential priorities, and potentially explosive promiscuity.

But the digital barrier against pandemics affects everyone and the social and health issues take on a forced global dimension, which cannot yield to myopic localisms. Even the regional fragmentation of healthcare shows evident limits even in our country because viruses have no residence, regardless of bureaucratic localisms that struggle to find a coordinated synthesis at the national and supranational levels.

Digital platforms are also the pivot of e-health or mobile-health (telemedicine) applications, which gradually spread, allowing in nonacute cases the domiciliation of patients, with enormous savings and an appreciable increase in the quality of life. It decongests hospitals and slows infections. The interdisciplinary scientific research of doctors, computer scientists, etc., can facilitate important advances in this area, geographically scalable around the world. Vaccines break the chain of infections, immunizing the nodes, and block the R_0 replication factors.

Digital platforms are a technology-enabled transactional tool that facilitates connections between stakeholders (the hospital as an organizational entity as well as doctors, nurses, other employees, patients, suppliers, financial lenders, etc.). Their features are consistent with network theory applications where stakeholders are the nodes rotating around the platform. Platforms may be considered a new virtual stakeholder that, consistently with network theory, connects conventional partners (shareholders, managers, employees, lenders, clients, suppliers, etc.), representing a bridging node and edge in multilayer networks. Stakeholders are nodes that interact around the bridging node, sharing information and cocreating value within a sustainable digital ecosystem. Shared information is fueled in real time by big data, and it reduces asymmetries and risk, redesigning information systems.

Platforms are facilitators of exchange (goods, services, and information) between different types of stakeholders that could not otherwise interact with each other. Transactions are mediated through complementary players that share a network ecosystem (Rochet and Tirole, 2003; Armstrong, 2006). Due to their digital features, they have a global outreach that gives them the potential to scale.

Digitalization is the process of converting data (not necessarily healthcare information) into a computer-readable format. Digital platforms are "software-based external platforms consisting of the extensible codebase of a software-based system that provides core functionality shared by the modules that interoperate with it and the interfaces through which they interoperate" (Tiwana et al., 2010). Software platforms are a technological meeting ground where application developers and end users converge (Evans et al., 2006).

Digitalization cannot be understood as an off-the-shelf product, bought as a one-time purchase in a warehouse. It rather requires a constantly developing vision that comes with a

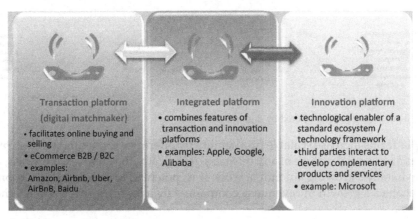

FIG. 3   Platform taxonomy.

continuous transformation process, hand in hand with strategic innovation management (Topol, 2019). Thus, digitalization means understanding the digital maturity level of an enterprise and the digital skills of the employees. Besides investment in products, a successful digitalization process also necessitates consideration of the cost to release employees from their obligations to contribute to the process as well as for a dedicated and continuing staff training and education program (Deiters et al., 2018).

A taxonomy of the platform typologies is shown in Fig. 3.

Digital platforms (Iyawa et al., 2017; Menvielle et al., 2017; Velez-Lapão, 2019; EXPH (Expert Panel on Effective Ways of Investing in Health), 2019) are increasingly used in healthcare (Ranerup et al., 2016), with declinations represented by e-health, m-health, or telemedicine.

Platforms reshape transaction costs among the value and supply chains, fostering information (big data mining; see Koh and Tan, 2011) sharing and smoothing transactions.

3.1 Networking effects and scalability of digital platforms

Networks become more valuable and important if they increase in size and interconnections, following Metcalfe's law according to which the effect of a telecommunications network is proportional to the square of the number of connected users of the system (n^2). Metcalfe's law (for critical analysis, see Odlyzko and Tilly, 2005) describes the network effects typical of communication technologies and networks such as the Internet and social networking. Metcalfe's law expresses mathematically the number of unique possible connections in a network of nodes. If a network is composed of n people and each of them assigns to the network a value proportional to the number of other participants, then the value that all n people assign to the network is:

$$n^* (n - 1) = n^2 - n \tag{1}$$

FIG. 4   Value-generating process ignited by digital platforms.

Digital platforms increase scalable profitability by offering exponential ecosystem growth. Scalability is an essential feature of any business. It indicates the ability of a process, network, or system to handle a growing amount of work or its potential to be enlarged to accommodate growth. It enables a growth in revenue accompanied by a less-than-proportional increase in variable costs.

Scalability is the ability of a device to adapt to a changing environment with changing customer needs. In broader terms, scalability means flexibility (so incorporating real options to expand, postpone, abandon businesses), which allows us to better address and achieve specific needs of clients with a customer-centric approach. People's interests and tastes as well as environmental conditions evolve continuously. Scalability is therefore vital as it contributes to competitiveness, efficiency, and quality (Moro Visconti, 2020). Scalability helps the system work without any delay or resource waste, making efficient use of available resources (Gupta et al., 2017).

The interaction among the stakeholders is eased by the digital platform bridging properties, fostering the incentive to cocreate and then share the additional value, as shown in Fig. 4.

Fig. 4, with its double-sided arrow, shows that value cocreation is a circular process because the involved stakeholders have joint incentives to foster digitalization to monetize and share its proceeds.

4. Introducing digital nodes: A network theory interpretation

The impact of digital platforms can be better understood considering their incremental function, confronted with an initial situation where digitalization (graphically represented by a bridging node) is absent.

Figs. 5 and 6 back a "without-with" comparison, showing—with a simplified ecosystem's wiring diagram—a standard healthcare network, and, respectively, a digital network mastered by a platform. The example illustrates the relationship between a public hospital that guarantees general healthcare coverage, and the private partners that supply ancillary functions (e.g., laundry, catering, accommodation, parking, ICT, etc.).

The private player is backed by subcontractors and the banks that support both the public and the private entities. The situation changes when a digital node is introduced, easing the exchange of information (big data) and transactions.

The compared analysis of Figs. 5 and 6 shows that in the latter, the digital platform acts as an intermediating (bridging) hub, increasing the number of nodes (vertices)–and so the overall value and consistency of the network—but especially the quantity and quality of the links.

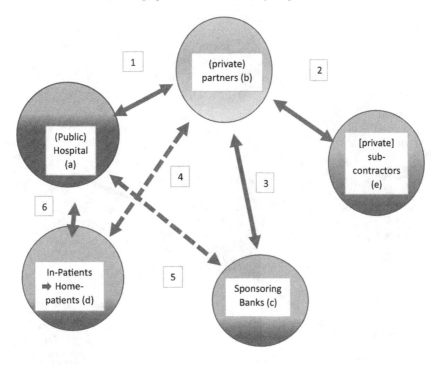

Legenda

1. Private to Public invoicing: private income (cash-inflow) and specular public costs (cash-outflows)

2. Sub-contractors to Private invoicing: private costs (cash-outflows) and sub-contractors income (cash-inflow)

3. Private to Bank negative interests (costs and cash-outflows) and specular bank to private positive income

  (Revenues and cash-inflows); bank to private financing and payback

4. Private supply to patients of non-core healthcare services

5. Treasury intermediation (public to private payments are mediated by the banking agent)

6. Public to the patient supply of services and patient to public payment of tickets

FIG. 5   Standard healthcare network.

For instance, any interaction between two agents mediated through the platform is digitally recorded in real time and may fuel big data gathering and artificial intelligence elaboration.

The added value embedded in Fig. 6 (compared to Fig. 5) can be interpreted with network theory analysis (Barabási, 2016), with a mathematical measurement of the degree of the nodes (number of links with other nodes), and a consequent estimate of their economic value. New connecting nodes with a system's centrality convey both information and economic transactions, partially reflecting the value increases of the PF parameters.

The value of each network can be estimated with Metcalfe's law, according to which the effect of a network is proportional to the square of the number of connected users (nodes) of the system (n^2). So, network$_{\text{figure 3}}$=25, and network$_{\text{figure 4}}$=36 (considering, for simplicity, that both networks have the same weights that measure the value of each link. This may underestimate the real value of the links of the platform).

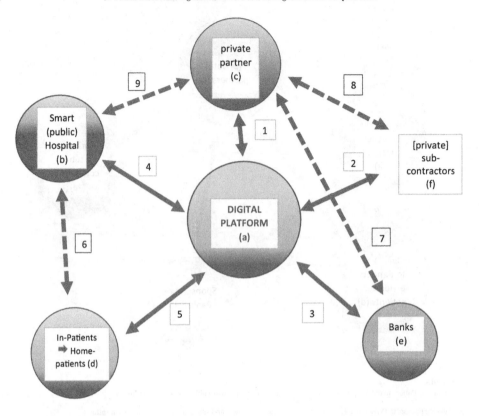

Legenda

1 + 4. Private to Public invoicing Through the Digital Platform: private income (cash-inflow) and specular public costs (cash-outflows)

2 + 1. Sub-contractors to Private invoicing through the Digital Platform: private costs (cash-outflows) and sub-contractors income (cash-inflow);
digital B2B auctions are conducted through the platform, with time and cost savings along the digitized supply chain.

3 + 1. Private to Bank negative interests (costs and cash-outflows) and specular bank to private positive income (revenues and cash-inflows) through the Digital Platform; bank to private financing and
payback

1+5. Private supply to patients (and visitors) of non-core healthcare services

4 + 5. Public to the patient supply of services and patient to public payment of tickets through the Digital Platform

5. Digital benefits for patients may be detected with Cost-Benefit Analysis and Cost-Effectiveness Analysis [Bergmo, 2015].

6. Direct contact between patients and hospital (for healthcare treatment, etc.).

7. Indirect relationship between the private actor and the sponsoring banks (not intermediated through the digital platform / mobile banking).

8. Physical supply of goods and services to the private actor.

9. Contractual public-to-private agreements. Physical supply of goods and services.

FIG. 6   Digital platform intermediating a smart healthcare network.

The digital platform, potentially operating always and everywhere, reduces paths and distances through its intermediating function that minimizes the number of links among the other nodes (shortest path). It also increases the network connectedness, creating additional paths between otherwise disconnected nodes (for instance, banks and subcontractors that are connected through the platform in Fig. 6 but are not connected in Fig. 5).

B. Enabling technologies for digital transformation (an example-oriented approach)

It is basically shown even in mathematical terms that the digitally mastered network outperforms the original (simple) network. The impact on bankability, although not directly calculated, is deemed to be positive. This also derives from the network robustness, fostered by the digital plasticity of resilient platforms.

A key property of each node is represented by its degree—the number of links it has to other nodes. Digital platforms are bridging nodes that represent a hub and an exchange point for the sharing of information (big data) and economic transactions.

Digitalization and the Internet reduce physical paths and distances with online timeless connectedness. An example is given by e-health (telemedicine) where the distance between two or more nodes (the physician, the patient, etc.) does not matter. Ubiquity, ignited by digital universality, softens topological diversity, fostering standard uniformity among complex and heterogeneous healthcare systems.

Within the healthcare system, big hospitals represent a physical cluster (hub) that attracts other smaller nodes (peripheral outpatient clinics, etc.). The edging links can be described with a network diagram. Following the preferential attachment rule, nodes prefer to link with more connected nodes, which in our case are mastered by digital platforms.

Healthcare networks are typically undirected, as stakeholders (physicians, patients, etc.) interact in both directions, being mastered by digital platforms that represent an IT interface for the exchange of data.

Resilience to risk is measured by network robustness against errors and failures. A sound IT architecture greatly matters, even considering the growing threat of cyberattacks. The network's robustness is measured by percolation theory, which considers the impact of the removal of a single or more nodes. Attack tolerance measures the potential impact of cascading failures (blackouts, information cascades, etc.) on the networked ecosystem. Building robustness is a forward-looking strategy that improves the resilience of the healthcare ecosystem, reducing criticalities.

4.1 Multilayer networks

A further extension of the network interpretation of Figs. 5 and 6 may be represented by multilayer networks (Bianconi, 2018) that are connected thanks to the presence of bridging digital platforms, as shown in Fig. 7.

Fig. 7 illustrates the centripetal impact of the digital platform on the layers that get closer and superimposed in some areas, sharing some common nodes, and intensifying intralayer and interlayer edges. Multilayer networking increases the overall value of the PPP ecosystem and may inspire frontier interdisciplinary research.

5. The innovative impact of scalable intangibles: From big data to artificial intelligence and blockchains

Scalable intangibles represent the backbone of e-health digital infrastructures. The flow that presides over this value chain is shown in Fig. 8.

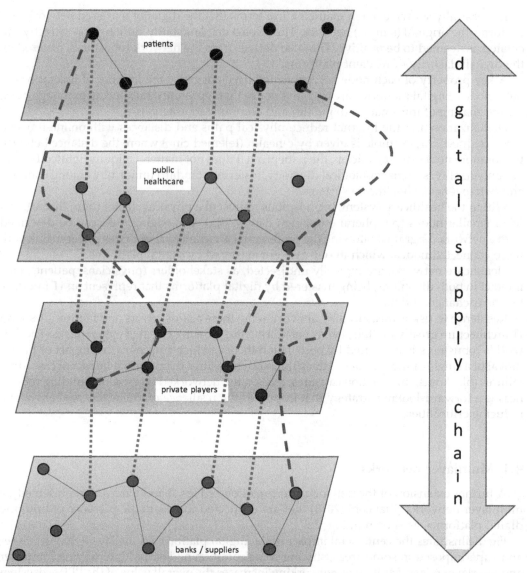

FIG. 7 Multilayer healthcare networks with a bridging digital platform.

FIG. 8 Scalable intangibles.

B. Enabling technologies for digital transformation (an example-oriented approach)

Big data in healthcare (Witjas-Paalberends et al., 2018) represent a precious source (often nurtured by Internet of Things (IoT) devices and wearables), to be validated with blockchains (Mazlan et al., 2020; Agbo et al., 2019; Gordon and Catalini, 2018) and interpreted with artificial intelligence analytics (Jiang et al., 2017). This sequential value chain increases the overall functionality of the digitally mastered healthcare supply chain, strengthening its resilience.

The value of information in healthcare has often been neglected or underreported and is now fully understood by digital processes that allow appropriate recording, storage, and interpretation.

Due to the impact of COVID-19) automation and artificial intelligence have attracted renewed interest in multiple industrial fields. Global manufacturing bases were affected strongly by workforce shortages associated with the spread of COVID-19. They are working to increase productivity by embracing digital manufacturing technologies that take advantage of artificial intelligence and the IoT (Javaid and Khan, 2021) that offer the promise of improved connectivity among supply chains (Kim and Lee, 2021).

6. Discussion

Sustaining economic activities while curbing the number of new COVID-19 cases until effective vaccines or treatments become available is a major public health and policy challenge (Nishi et al., 2020).

This chapter shows how digital platforms, mapped and interpreted with network theory analysis, can be used for an innovative interpretation of smart healthcare systems (Deiters et al., 2018).

Digitalization is changing businesses everywhere and the healthcare sector is no exception. Yet history teaches us that radical innovation rarely comes from within the industry (Vermeer and Thomas, 2020). There are important implications for the digital transformation vision in healthcare that can be improved using network theory and its mathematical properties to map the interactions among complementary stakeholders. Alignment of interests and minimization of conflicts is a byproduct of a reengineered ecosystem where digitalization eases networking, acting as a real-time catalyzer.

Digitalization is, however, still in its infancy, and practical applications concerning healthcare must face enormous challenges. An example is the lack of interoperability among databases where spurious data are often randomly collected and stored. Further research and on-field consultancy are needed to smooth and overcome these difficulties.

Empirical evidence on health services throughout the COVID-19 pandemic has been scarcely documented in the literature (Davis, 2020). Specific research may fill this gap, providing professional advice for the future where similar cases are likely to occur.

Healthcare represents a large and hopelessly complicated ecosystem that can be innovatively interpreted with network analysis. If we want to understand a complex system, we first need to know how its components interact with each other (Barabási, 2016, p. 45). Digital platforms reduce the entropy of disordered systems, increasing virtual connectedness and accelerating growth, so fostering telemedicine applications.

Kinetic analysis of evolutionary networks describes their changes over time, incorporating technological advances that are fostered by emergencies such as the pandemic. Degree dynamics show each node's temporary evolution. All these literature streams, well known in network theory, deserve empirical applications in the healthcare industry.

Policy recommendations for both academics and practitioners include increased awareness of networking properties fostered by digital platforms that master the ecosystem, smoothing its working and increasing its resilience and efficiency. To the extent that healthcare is an intrinsically highly networked industry, digital nodes (platforms) can substantially contribute to its improvement. Practical applications are also catalyzed by technological upgrades, where "physical" devices (e.g., diagnostic equipment) are complemented by software and other intangibles to make networking instantaneous and ubiquitous.

7. Conclusion

This study analyzes the impact of smart healthcare digitalization during the COVID-19 pandemic. This type of setting could have important contributions to the research and practice of healthcare trends, consistently with the sustainable development goal 3 (ensure healthy lives and promote well-being for all at all ages), and with broader environmental, social and governance (ESG) targets that measure ecology-compliant socioeconomic sustainability. ESG consciousness goes beyond the empty promise of never-ending growth that is embedded in old-style capitalism.

Environmentally friendly innovation, levered by digitalization and other intangibles, shapes a new paradigm around sustainable capital. As Alan Eddision warns, modern technology owes ecology an apology.

This chapter shows that digital platforms exemplify a bridging node that reshapes the networked interaction of connected healthcare stakeholders, easing social distancing and favoring remote diagnosis.

In its industry application, the research concentrates on healthcare investments that concern an essential social infrastructure characterized by increasing sustainability issues, mainly because of public budgetary pressures and the aging population (Yang et al., 2021).

There are manyfold practical implications of this study.

First, digitalization adds value, with a positive impact on financial and economic marginality. This is magnified both by the long-term schedule of healthcare investments and by the still undetected likelihood that innovation (concerning the synergistic interaction of big data, cloud computing, artificial intelligence, and blockchains) could boost productivity in this data- and technology-sensitive industry.

Second, value cocreation can reward all stakeholders, improving the value for money but also the healthcare performance in a sector where quality can make the difference between survival and death. The benefit extracted from the sharing of the value "pie" (with results-based financing, pay-for-performance, or other rewarding agreements) represents an incentive for all stakeholders to nurture a self-fulfilling value cocreation process, consistent with the representation of Fig. 7.

Disruptive technological advances will also contribute to reshaping the digital supply and value chain, shortening its passages and producing transactional savings. New research streams may be inspired by oncoming evidence of the digital impact on business modeling and the value creation of innovative healthcare investment patterns. Digital platforms may contribute to reshaping the life cycle of healthcare projects, improving their resilience and softening the risk concerns that endanger long-term economic sustainability.

The findings of this study may be extended to industries outside healthcare or generalized to other stakeholder connections bridged by digital platforms and interpreted with network theory principles. Patient-centric stakeholdership issues (Moro Visconti and Martiniello, 2019) deserve increasing attention, representing further research directions.

The paradox that emerges from network theory is that in their multivariate applications, they look like a double-edged sword: network links through pathogens are problematic in epidemics but virtual links in e-health applications can represent an aseptic bridge that prevents contagion and fosters many healthcare applications. These interactions have seldom been investigated, but they represent a trendy issue for future interdisciplinary research. Health experts (medical doctors, computer scientists, policymakers, government officers, educators, big pharma and MedTech managers, etc.), well aware of the contagion effect, may find innovative inspiration in its mathematical and IT properties based on networking digital platforms that can be mapped and programmed for efficiency scale-up. Patients, on their side, evolve beyond a passive acquaintance and have a say, increasingly interacting with other stakeholders.

Interdisciplinary research, like the one exemplified in this chapter, is vital for practical improvements in this multifaceted environment where technology adapts to medical issues, showing value cocreation patterns that eventually bring win-win pathbreaking solutions.

Transformation of in patients into daily out patients and, eventually, home patients (Moro Visconti et al., 2020) represents, whenever possible (e.g., in the absence of acute treatment), a mighty goal for smart healthcare upgrades, catalyzed by aseptic digital networking.

Appendix

Data deposition: Moro-Visconti, Roberto (2021), Impact of healthcare digital platforms, Mendeley Data, V1, https://doi.org/10.17632/zxdp38mv42.1.

Suggested teaching assignments

Build up an empirical case of a telemedicine intervention, analyzing with a questionnaire the diffusion of an M-App for the distant monitoring of chronic disease (e.g., diabetes).

Recommended readings

See References.

B. Enabling technologies for digital transformation (an example-oriented approach)

Recommended videos

World Economic Forum, Shift to digital during the pandemic could enable universal health coverage, https://www.weforum.org/agenda/2020/12/shift-to-digital-during-pandemic-could-enable-universal-health-coverage/.
GovTech Singapore: Responding to COVID-19 with Tech, https://www.tech.gov.sg/products-and-services/responding-to-covid-19-with-tech/.

Titles for research essays

Validating healthcare digital networks with blockchains
Interpretation of e-health with artificial intelligence
Big data, IoT, cloud computing, and information sharing in healthcare value chains
Mathematical and statistical modeling of healthcare networks, concerning epidemics or digital platform optimization
Cost/benefit analysis and pay-for-performance screening of digital applications in healthcare

Recommended projects URL

www.morovisconti.com/en
https://www.researchgate.net/profile/Roberto-Moro-Visconti

References

Agbo, C., Mahmoud, Q., Eklund, J., 2019. Blockchain technology in healthcare: a systematic review. Healthcare 7, 56.
Armstrong, M., 2006. Competition in two-sided markets. RAND J. Econ. 37 (3), 668–691.
Asadullah, A., Faik, I., Kankanhalli, A., 2018. Digital platforms: a review and future directions. In: Twenty-Second Pacific Asia Conference on Information Systems, Japan. https://www.academia.edu/37873177/Digital_Platforms_A_Review_and_Future_Directions.
Barabási, A., 2016. Network Science. Cambridge University Press, Cambridge.
Baumgart, D.C., 2020. Digital advantage in the COVID-19 response: perspective from Canada's largest integrated digitalized healthcare system. NPJ Digit. Med. 3, 114.
Bianconi, G., 2018. Multilayer Networks. Oxford University Press, Oxford.
Bokolo Jr., A., 2020. Use of telemedicine and virtual care for remote treatment in response to COVID-19 pandemic. J. Med. Syst. 44, 132.
Cipolla, C.M., 1981. Faith, Reason, and the Plague in Seventeenth-Century Tuscany. W.W. Norton & Company.
Davahli, M.R., Fiok, K., Karwowski, W., Aljuaid, A.M., Taiar, R., 2021. Predicting the dynamics of the COVID-19 pandemic in the United States using graph theory-based neural networks. Int. J. Environ. Res. Public Health 18, 3834.
Davis, R., 2020. Integrating digital technologies and data driven telemedicine into smart healthcare during the COVID-19 pandemic. Am. J. Med. Res. 2, 22–28.
Deiters, W., Burmann, A., Meister, S., 2018. Strategies for digitalizing the hospital of the future. Urologe A 57 (9), 1031–1039.
Dossi, F., Marini, A., Sanna, G.D., Saba, P.S., Parodi, G., 2020. M-health and telemedicine: are people ready for the future? A real-world survey. Eur. Heart J. 41 (Suppl_2).

Evans, D.S., Hagiu, A., Schmalensee, R., 2006. Invisible Engines. How Software Platforms Drive Innovation and Transform Industries. MIT University Press, Cambridge.

EXPH (Expert Panel on Effective Ways of Investing in Health), 2019. Assessing the Impact of Digital Transformation of Health Services. https://ec.europa.eu/health/sites/health/files/expert_panel/docs/022_digitaltransformation_en.pdf.

Frith, J., 2012. The history of plague—part 1. The three great pandemics. History 20 (2).

Gastaldi, L., Corso, M., 2012. Smart healthcare digitalization: using ICT to effectively balance exploration and exploitation within hospitals. Int. J. Eng. Business Manag. 4.

Golinelli, D., Boetto, E., Carullo, G., Nuzzolese, A.G., Landini, M.P., Fantini, M.P., 2020. Adoption of digital technologies in health care during the COVID-19 pandemic: systematic review of early scientific literature. J. Med. Internet Res. 22 (11).

Gómez, J., Verdú, M., 2017. Network theory may explain the vulnerability of medieval human settlements to the Black Death pandemic. Sci. Rep. 7, 43467.

Gordon, W.J., Catalini, C., 2018. Blockchain technology for healthcare: facilitating the transition to patient-driven interoperability. Comput. Struct. Biotechnol. J. 16, 224–230.

Gupta, A., Christie, R., Manjula, R., 2017. Scalability in internet of things: features, techniques and research challenges. Int. J. Comput. Intell. Res. 13 (7), 1617–1627.

Haleem, A., Javaid, M., 2020. Medical 4.0 and its role in healthcare during COVID-19 pandemic: a review. J. Indus. Integr. Manag. 5 (4), 531–545.

Harari, Y.N., 2015. Sapiens: A Brief History of Humankind. Harper, New York.

Herrmann, M., Boehme, P., Mondritzki, T., et al., 2018. Digital transformation and disruption of the health care sector: internet-based observational study. J. Med. Internet Res. 20 (3), e104.

Iyawa, G.E., Herselman, M., Botha, A., 2017. A scoping review of digital health innovation ecosystems in developed and developing countries. In: Conference: IST Africa, At Windhoek, Namibia. https://www.researchgate.net/publication/318013312_A_Scoping_Review_of_Digital_Health_Innovation_Ecosystems_in_Developed_and_Developing_Countries.

Javaid, M., Khan, I.H., 2021. Internet of Things (IoT) enabled healthcare helps to take the challenges of COVID-19 Pandemic. J. Oral Biol. Craniofacial Res. 11 (2), 209–214.

Jiang, F., Jiang, Y., Zhi, H., Dong, Y., Li, H., Ma, S., Wang, Y., Dong, Q., Shen, H., Wang, Y., 2017. Artificial intelligence in healthcare: past, present, and future. Stroke Vascular Neurol. 2, 230–243.

Kalhori, S.R.N., et al., 2021. Digital health solutions to control the COVID-19 pandemic in countries with high disease prevalence: literature review. J. Med. Internet Res. 23 (3), e19473.

Kelly, J., 2005. The Great Mortality: An Intimate History of the Black Death, the Most Devastating Plague of All Time. Penguin, London.

Kim, H.K., Lee, C.W., 2021. Relationships among healthcare digitalization, social capital, and supply chain performance in the healthcare manufacturing industry. Int. J. Environ. Res. Public Health 18 (4), 1417.

Koh, H.C., Tan, G., 2011. Data mining applications in healthcare. J. Healthc. Inform. Manag. 19, 64–72.

Kraus, S., Schiavone, F., Pluzhnikova, A., Invernizzi, A.C., 2021. Digital transformation in healthcare: analyzing the current state-of-research. J. Bus. Res. 123, 557–567.

Mazlan, A.A., Daud, S.M., Yusof, M.F., et al., 2020. Scalability challenges in healthcare blockchain system—a systematic review. IEEE Access 8.

Menvielle, L., Audrain-Pontevia, A., Menvielle, W., 2017. The Digitization of Healthcare. New Challenges and Opportunities. Palgrave Macmillan, Cham.

Moro Visconti, R., 2020. The Valuation of Digital Intangibles. Technology, Marketing and Internet. Palgrave-Macmillan, Cham.

Moro Visconti, R., Larocca, A., Marconi, M., 2020. Accessibility to First-Mile health services: a time-cost model for rural Uganda. Soc. Sci. Med. 265, 1–12.

Moro Visconti, R., Martiniello, L., 2019. Smart hospitals and patient-centered governance. Corp. Ownersh. Control. 16, 83–96.

Nagurney, A., 2021. Supply chain game theory network modeling under labor constraints: applications to the Covid-19 pandemic. Eur. J. Oper. Res. 293 (3), 880–891.

Nishi, A., et al., 2020. Network interventions for managing the COVID-19 pandemic and sustaining economy. Proc. Natl. Acad. Sci. 117 (48), 30285–30294.

Odlyzko, A., Tilly, B., 2005. A Refutation of Metcalfe's Law and a Better Estimate for the Value of Networks and Network Interconnections. University of Minnesota, Minneapolis.

B. Enabling technologies for digital transformation (an example-oriented approach)

Ranerup, A., Zinner Henriksen, H., Hedman, J., 2016. An analysis of business models in public service platforms. Gov. Inf. Q. 33, 6–14.

Reale, G., 1990. A History of Ancient Philosophy II—Plato and Aristotle. State University of New York Series in Philosophy, New York.

Rochet, J.C., Tirole, J., 2003. Platform competition in two-sided markets. J. Eur. Econ. Assoc. 1 (4), 990–1029.

So, M.K.P., Chu, A.M.Y., Tiwari, A., Chan, J.N.L., 2021. On topological properties of COVID-19: predicting and assessing pandemic risk with network statistics. Sci. Rep. 11, 5112.

So, M.K.P., Tiwari, A., Chu, A.M.Y., Tsang, J.T.Y., Chan, J.N.L., 2020. Visualizing COVID-19 pandemic risk through network connectedness. Int. J. Infect. Dis. 96, 558–561.

Sutherland, W., Jarrahi, M.H., 2018. The sharing economy and digital platforms: a review and research agenda. Int. J. Inf. Manag. 43, 328–341.

Swanson, D., Santamaria, L., 2021. Pandemic supply chain research: a structured literature review and bibliometric network analysis. Logistics 5 (1), 7.

Tiwana, A., Konsynsky, B., Bush, A.A., 2010. Platform evolution: coevolution of platform architecture, governance, and environmental dynamics. Inf. Syst. Res. 21 (4), 675–687.

Topol, S., 2019. A decade of digital medicine innovation. Sci. Transl. Med. X (11).

Velez-Lapão, L., 2019. The challenge of digital transformation in public health in Europe? Eur. J. Pub. Health 29 (4).

Vermeer, L., Thomas, M., 2020. Pharmaceutical/high-tech alliances; transforming healthcare? Digitalization in the healthcare industry. In: Strategic Direction.

Witjas-Paalberends, E.R., van Laarhoven, L.P.M., van de Burgwal, L.H.M., Feilzer, J., de Swart, J., Claassen, E., Jansen, W.T.M., 2018. Challenges and best practices for big data-driven healthcare innovations conducted by profit-non-profit partnerships—a quantitative prioritization. Int. J. Healthc. Manag. 11, 171–181.

Yang, Y., Zheng, R., Zhao, L., 2021. Population aging, health investment and economic growth: Based on a cross-country panel data analysis. Int. J. Environ. Res. Public Health 18, 1801.

Factors influencing the adoption of mobile health apps in the UAE

Haseena Al Katheeri[a], Nazia Shehzad[b], and Fauzia Jabeen[c]
[a]Center for Student Success, Zayed University, Abu Dhabi, United Arab Emirates [b]Faculty of Business, Liwa College of Technology, Abu Dhabi, United Arab Emirates [c]College of Business Administration, Abu Dhabi University, Abu Dhabi, United Arab Emirates

1. Introduction

The recent developments in information technology and the use of smart devices have revolutionized the mobile application industry. Firms have already felt the need to embark on a digital transformation journey to sustain themselves (Chatterjee et al., 2022). Almost every sector has or is in the process of adopting mobile applications to reach their intended users, from education, science, products, and service industries to the military and health sectors. One of the areas that has significantly benefited from these advancements is the health sector (Azadi et al., 2023). Mobile technology use for healthcare has been advanced through mobile health (mHealth) applications (apps). These digital health platforms have enabled medical care access and service from the comfort of homes or other places providing Internet access.

There are different platforms for mHealth, including Apple iOS and Android (Götze, 2015). mHealth apps such as FitBit, Map My Fitness, Jawbone (Gao et al., 2015; Götze, 2015), and many others are widely used, with more than 3.7 billion downloads over the past 5 years (Statista, 2018). mHealth is considered a disruptive innovation that employs a wireless mobile network to aid health service delivery through sensors attached to the body or directly through the mobile device itself (Deterding et al., 2011; Wen, 2017). Various users adopt these applications because they make them aware of their health through self-monitoring. They track heartbeats, the number of steps, burned calories, sleep hours, weight measurements, water intake, and other general well-being activities (Thirumurthy and Lester, 2012).

Technological advances manifested in approaches such as deep learning and artificial intelligence (AI) are of substantial importance throughout healthcare services, and are understood as competitive advantages (Alkatheeri et al., 2021; Azadi et al., 2023). The ubiquitous nature of mHealth technologies motivates research to study the potential use and adoption of these technologies by consumers and patients to maintain their health. One of the medical areas that these technologies has impacted is the management of chronic diseases. These are persistent and incurable medical conditions caused by genetic transfers or the adoption of unhealthy habits such as tobacco use, poor diets, or lack of physical exercise. Some chronic medical conditions include hypertension, diabetes, cancer, obesity, and cardiovascular diseases such as hypercholesterolemia caused by unhealthy diets. If not managed effectively, these conditions are highly fatal. Many applications have come in handy in managing and controlling the severity of these diseases. Most of these conditions can be controlled through physical exercises, dieting, and medication (Steinhubl et al., 2015). This has led to the creation of medical applications for dieting and physical wellness to help users schedule and routinely undertake exercise and monitor their food consumption to ensure a healthy diet to control and manage the severity of the diseases. However, despite the high demand and download rates as well as the rise in mHealth app market value to $4.9 billion, the severity of chronic disease conditions has persisted. Recent statistics show increased use of mHealth applications, but little impact on patient and user health has been documented. This has been the worrying trend that necessitated this research.

The usability of these applications is influenced by many factors, such as the interface and design (Mendiola et al., 2015). Most mHealth apps have a complex and unfriendly user interface, making it hard for users to navigate through the numerous segments and information. Therefore, once downloaded, most of these applications lie idle in the smartphone memory because they are complex and lack other uses (Zahra et al., 2017). Furthermore, there are few confidence and security assurances on user data (Dehling et al., 2015). There is sensitive user medical information that is entered during registration. The minimal assurances that the data uploaded are safely and confidentially stored have kept many users from these applications due to the trauma caused by stigmatization in case such personal information lands in the wrong hands (Zahra et al., 2017). Prior studies (Al Badi et al., 2022) also revealed that accuracy, privacy, and security criteria are the most important factors to optimize the healthcare sector with technology.

Getting these applications is easy because they are readily available in app stores. However, using them has been the most challenging part (Scheibe et al., 2015). People need to be more motivated when it comes to exercising. In most cases, people want to be pushed or monitored to adhere to the exercise routines stipulated. This has been quite a challenge. Most people are lazy and therefore need to use the applications more effectively. Some apps also require technical training and physical guidance, which many people cannot access.

mHealth apps are expected to provide effective and cost-efficient healthcare, which is often a concern for consumers (Nieroda et al., 2015). Thus, it is vital to assess the effects of this technology, which has revolutionized the way consumers move from traditional healthcare services to mHealth. This research focuses on behavior change when using mHealth apps. In this context, the Unified Theory of Acceptance and Use of Technology (UTAUT) model (Venkatesh et al., 2003) could facilitate adoption patterns of mHealth technology. The constructs of UTAUT, including its four primary constructs—performance expectancy (PE), effort expectancy (EE), social influence (SI), and facilitating conditions (FC)—are being used to investigate the intention toward the actual use (Venkatesh et al., 2003) of mHealth apps to improve self-managing of healthcare. In addition, this paper adapts gamification impact (GI) because it is common to use one or more gamified features in designing mHealth applications (Lister et al., 2014) to increase the practical use of these apps (Dennis and O'Toole, 2014). This study aims to analyze the different drivers of mHealth app behavior use and to validate a new theoretical framework, UTAUT (Venkatesh et al., 2003), in this context.

Prior studies recommended conducting this research in various cultures. Hence, it is helpful to apply it in the United Arab Emirates (UAE) due to its unique cultural characteristics as an emerging Arab country. Therefore, this research uses a theoretical framework to examine the factors influencing the acceptance of mHealth technology by using the UTAUT model combined with gamification impact in the UAE. Hence, this study aims to determine consumer readiness to use mHealth from a usability standpoint. This study seeks to answer the following research questions:

(a) What are the factors instrumental in influencing users' behavior to use mHealth apps?
(b) What is the role of personal innovativeness in influencing users' intention to use mHealth apps?

The remainder of this paper is organized as follows: The next section reviews relevant literature, followed by hypotheses development and the proposed research model for this study. The third section presents the research method, followed by data analysis and discussion in the fourth section. The last section discusses limitations and areas for further study.

2. Literature review

2.1 User acceptance of mHealth apps

Mobile health utilizes technology, such as smartphones, to observe and improve the health quality of users not only in developed countries but also in developing countries (Lewis et al., 2012). The mHealth marketplace is significant, estimated at $2.4 billion in 2013 and estimated to grow to $26 billion by 2017 (Miller, 2014). According to Fox and Duggan (2012), approximately one in five smartphone users utilize at least one mHealth app to boost their goals related to health, and 38% of health application users have downloaded at least one app for physical activity (PA). There is an accelerated and consistent growth in using mHealth apps, where statistics report that the estimated number of mHealth app downloads reached 3.7 billion worldwide in 2017 (Statista, 2018). Similarly, mobile apps were expected to be downloaded more than 268 billion times in 2017, generating revenue of more than $77 billion and making apps one of the most popular computing tools for users across the globe (Gartner, 2014). Additionally, more than 97,000 health-related apps are available in the health and fitness category of the Apple app store and Google play store, with about 1000 more being created every month.

mHealth apps serve various purposes, such as helping with smoking cessation and weight loss programs, monitoring diet and physical activity, and helping with medical treatments and disease management (*Measuring the Information Society*, 2012). The main advantages of using mobile phones and similar technologies for monitoring health are that they are personal, intelligent, connected, and always with the people (Fogg and Adler, 2009). Even though there is a need for more clarity on how consumers engage with and continue to use digital products for health, self-management remains a significant challenge for healthcare companies in developing effective digital strategies (Hird et al., 2016).

The UTAUT model (Venkatesh et al., 2003) is considered the most recent technology acceptance model used in research. It aims to identify the intention of individuals to use information systems through a unified view. The four main determinants of the UTAUT model are performance expectancy, effort expectancy, social influence, and facilitating conditions used to explain user acceptance and usage behavior.

There are various domains of technology implemented in adoption contexts using the UTAUT model, such as e-books (Maduku, 2015), e-banking, e-government services (AlAwadhi and Morris, 2008), document workflow management systems (Mosweu et al., 2016), healthcare robots (Alaiad et al., 2014), and mobile health services (Dwivedi et al., 2016). This is due to the robustness, comprehensiveness, statistical validity, reliability, and accuracy of the UTAUT model to predict technology acceptance in different disciplines using technology in various contexts (Maduku, 2015). Because the UTAUT model is excellent and empirically tested to study technology acceptance compared to other models (Park et al., 2007), the author assumes that this model can be used to understand users' behavioral intention toward using mHealth in the UAE for effective self-monitoring to improve health.

Prior studies in mHealth have used several fragmented theories associated with acceptance behavior, mainly theory of planned behavior (TPB), theory of reasoned action (TRA), technology acceptance model (TAM), and innovation diffusion theory (IDT). For instance, a study was made on the acceptance and adoption of smartwatches to be used in various activities, including fitness, health monitoring, and location tracking (Kim and Shin, 2015). Four constructs from TAM were used, which are perceived ease of use (PEOU), perceived usefulness (PU), attitude (AT), and intention to use (IU). Euehun and Semi (2015) only applied usefulness value from the TAM model in their study with other factors such as perceived needs (illness experiences), convenience value, and monetary value (Euehun and Semi, 2015). They reported that users changing behavior and intentions are determined by the usefulness value of mHealth (Euehun and Semi, 2015). Also, other researchers extended the TAM model in the healthcare context to comprehensively predict the adoption of healthcare technologies by various consumers (Euehun and Semi, 2015; Hossain, 2016; Kim and Shin, 2015; Lee et al., 2015, 2017; Park et al., 2016).

This theory represents a more integrative and current framework than the TAM and TPB models (Kaba and Touré, 2014; San Martín and Herrero, 2012b). Several researchers (Dwivedi et al., 2016; Mallat, 2007) asserted that there are issues in TAM, TBP, and IDT to be used in predicting consumer use behavior for online-based products. Due to these issues, the UTAUT model and two more constructs, Gamification Impact and Trust, have been used for the adoption of mHealth by users in the UAE.

2.2 Gamification in the field of mHealth

Gamification is "the utilization of game-design elements and principles in nongame contexts" (Lister et al., 2014). It offers motivational affordance in various tasks that support behavioral outcomes (Koivisto and Hamari, 2014). Gamification utilizes gamified dynamics in a nongaming context to invoke motivational affordance and psychological or behavioral outcomes (Lee et al., 2017; Zhao et al., 2016). It consists of intrinsic and extrinsic rewards (Lee et al., 2017) that encourage behavior through motivation, engagement, and the learning process (Filsecker and Hickey, 2014). There are eight elements of gamification: goal setting, capacity to overcome challenges, feedback, reinforcement, compare progress, social connectivity, playfulness, and self-monitoring (Kappen and Orji, 2017). Hence, gamification can be defined as using one or more game elements (Edwards et al., 2016) to foster changes in an individual's behavior and add enjoyable features to the task (Kappen and Orji, 2017).

Gamification is connected to a rewards system where most applications emphasize "adding points, levels, leaderboards, achievements or badges to a real-world setting" to make people engage with the activity to obtain their rewards (Nicholson, 2015). One of the most valuable elements for changing behavior in health technology is an intrinsic motivation to employ gamification effectively. Intrinsic motivation is based on the self-determination theory (Deci and Ryan, 2004), which combines three psychological needs: competence, autonomy, and relatedness (Nicholson, 2015). This theory is also grounded in making people less engaged in the intended behavior because they feel dull in gamified environments (Burke, 2014). Gamification creates a fun environment and enjoyment when playing a game using mechanics and dynamics to achieve user participation and engagement (Richter et al., 2015).

mHealth applications employ gamified features to increase the practical use of these applications (Dennis and O'Toole, 2014). The study by Lister et al. (2014) represented gamification use in mHealth, which became common in many technologies by implementing one or more components. The association of gamification with health behavior theory only considers the motivation element because the industry is trying to increase the motivation of individuals (Lister et al., 2014). On the other hand, Edwards et al. (2016) reported that health applications rarely employ gamification features while behavior change techniques are widely used.

2.3 Application of UTAUT and gamification

The validity of the UTAUT model in studying technology acceptance in many fields and the utilization of gamified components are available on most system designs. It is found that there is a positive impact of gamification in studying the behavior change of consumers when using mHealth. Therefore, in this research, the UTAUT model will be combined with the gamification impact component to analyze every instance of mHealth adoption and implementation and to increase the credibility and effectiveness of the study. This research integrates four main constructs in the UTAUT model with gamification principles.

Based on these findings, it can be concluded that most research efforts focused on using mHealth apps from the patient perspectives. To the best of our knowledge, studies have yet to be conducted from the actual use perspectives considering the effects of gamified features and personal innovativeness, even though the later determinants play significant roles in successfully adopting mHealth. In addition, most previous studies have been conducted in the context of developed nations. Therefore, this study attempted to fill these gaps by investigating the issues associated with the adoption and acceptance of mHealth apps in the context of UAE from the user perspectives. A summary of constructs from the UTAUT model and gamification is provided in Table 1.

TABLE 1 Literature synthesis on technology acceptance integrated with gamification.

Author/year	Subject	Sample	Theatrical/ conceptual model(s)	Factors affect usability
San Martín and Herrero (2012a)	Intention to purchase online	1083 valid surveys were acquired	UTAUT model	Performance expectancy, effort expectancy, social influence, facilitating conditions
Kaba and Touré (2014)	Behavioral intention to use and adopt social networking site	Users from 27 high schools and four universities	UTAUT	Performance expectancy, effort expectancy, social influence, facilitating conditions
Rodrigues et al. (2014)	Intention to use e-banking systems	219 e-banking customers	TAM, gamification	Web design, ease of use, gamification, information, webpage characteristic
Maduku (2015)	Behavioral intention of students to use e-books	439 students from five tertiary institutions (two public universities and three independent tertiary institutions)	UTAUT gamified features applied on e-books	Performance expectancy, effort expectancy, social influence, facilitating conditions
Almuraqab and Jasimuddin, (2017)	Intention to use m-government	83 respondents	TAM model	Perceived ease of use, perceived usefulness, social influence, trust in technology, trust in government, perceived compatibility
Baptista and Oliveira (2017)	Mobile banking services that use gamification	50 Brazilian customers	UTAUT and UTAUT2	Performance expectancy, effort expectancy, social influence, facilitating conditions, hedonic motivation, price value, habit, and gamification impact
Kumar and Acharjya, (2017)	Behavioral intention toward mobile games	50 participants who play games	UTAUT, Information systems success model	Visual effect of the game, effort expectancy, social influence, facilitating conditions, game system quality, satisfaction of the game

3. Hypotheses development

3.1 Performance expectancy (PE)

The concept of performance expectancy is based on the work of Venkatesh et al. (2003). Their studies demonstrated the importance of this variable for behavior intention from the perceived usefulness (TAM/TAM2) and relative management (IDT). In brief, performance expectancy can be defined as "the degree to which an individual believes that using the system will help him or her to attain gains in job performance" (Venkatesh et al., 2003).

In a study by Dwivedi et al. (2016), the importance of this variable is highlighted by its ability to predict individuals' thoughts to use mHealth for behavior change by defining it as judgment, meaning the extent that a well-performed behavior augments performance in a self-monitoring activity. Venkatesh et al. (2003) identified performance expectancy as the top determinant in predicting the behavioral intention and use of information technology.

Nevertheless, the UTAUT predictors had different weights dependent on the subgroup studied and thus suggested that every type of person should be treated separately (Devolder et al., 2012). It was also found that there was little evidence of the effect of performance expectancy and behavioral intention or actual use (Schaper and Pervan, 2006). Hence, we propose the following:

Hypothesis 1 Performance expectancy is positively related to behavioral intention.

3.2 Effort expectancy (EE)

Effort expectancy is defined as the extent of ease related to the use of technology (Venkatesh et al., 2003). Similar to performance expectancy, effort expectancy is extracted from its foundation from perceived ease of use (TAM/TAM2) and IDT (ease of use). In studies executed around the world, the technologies are diverse, including e-books (Maduku, 2015), social networking sites (Kaba and Touré, 2014), and electronic banking systems (Baptista and Oliveira, 2017). Effort expectancy was hypothesized to predict the acceptance of mobile healthcare service systems by citizens and evidence was found for this assumption in all three countries: the United States, Canada, and Bangladesh (Dwivedi et al., 2016).

Venkatesh et al. (2003) understand that effort expectancy refers to the amount of effort required to use a presented tool, in this case, mHealth technologies, or the user's understanding that applying such technologies will reduce the amount of physical and mental work. Therefore, the author presents the next hypothesis, which, based on Hypothesis 2, relates to a personal understanding of how behavior can change by understanding the benefits of new behavior, linking perceived consumer effectiveness to effort expectancy. Hence, by understanding how to use the technology, the user must also feel that the effort to use new technology is reduced. Although most researchers supported this relationship (Maduku, 2015; San Martín and Herrero, 2012a), others asserted that effort expectancy had no significant influence (Arman and Hartati, 2015; Bennani and Oumlil, 2013). Therefore, we propose the following:

Hypothesis 2 Effort expectancy is positively related to behavioral intention.

3.3 Social influence (SI)

According to Venkatesh et al. (2003), social influence is described as the belief that a person is affected by other people to accept that he should use a specific technology. These people might be family members, friends, colleagues, or managers in the workplace, followers on social media, or others. Due to these reasons, the person will be convinced to use this technology, and people will encourage him to do so. It is notable that this determinant strongly impacts the individual behavioral intention. Additionally, social influence is a major factor in behavioral motivation to accept mHealth, supported by the influence of the reference group (Dwivedi et al., 2016). So, there is no relationship between subjective norms or social influence and intention to use technology (Davis et al., 1989). Venkatesh et al. (2003) reported that there is a considerable effect of subjective norms under constant conditions. In this research, the author seeks to understand the social influence on behavioral intention to use mHealth applications; therefore, considering these aspects, the author proposes that social influence positively affects behavioral intention.

However, some researchers reported that social influence had only a limited effect on behavioral intention (Chang et al., 2007). The opposite study declared that this variable is considered the most salient predictor (Alaiad and Zhou, 2014). Furthermore, some studies rejected the hypothesis because of the lack of significant effect of social influence on the intention to use technology (Bennani and Oumlil, 2013; Phichitchaisopa and Naenna, 2013; San Martín and Herrero, 2012b). There are two possible explanations for this incident: first, the changes of timing in the study are just instituted, and second, some respondents who had more self-confidence and experience were less influenced by social pressure (Chang et al., 2007; Phichitchaisopa and Naenna, 2013).

Hypothesis 3 Effort expectancy is positively related to behavioral intention.

3.4 Facilitating conditions (FC)

One of the main UTAUT model factors is facilitating conditions, which are known as "the degree to which an individual believes that an organizational and technical infrastructure exists to support the use of the system." (Venkatesh et al., 2003). It is critical to producing continuous communication between the provider of mHealth service and the end user in different locations (Dwivedi et al., 2016). The resources and functions of the service provider should effectively provide the appropriate service in a timely and trusted manner, and this is supported by the user's conviction to adopt any technology regardless of the location or time (Peekhaus, 2008; Shareef et al., 2013; Weerakkody et al., 2011). When the user compares traditional healthcare services and mHealth applications, he will consider security, privacy, and reliability (Gefen et al., 2003; Mallat, 2007; Pavlou and Fygenson, 2006). In this sense, the variable of facilitating conditions has played a sensitive role in determining the behavioral intention of consumers to use the mHealth application.

Researchers found the facilitating conditions predictor was the most salient predictor; others found that it had a marginally significant effect (Dajani and Yaseen, 2016). Indeed, some researchers even excluded this variable from their investigations (Abushanab and Pearson, 2007; Koivumäki et al., 2008). In comparison, others propose that the impact of facilitating conditions is directly related to the behavior use instead of acceptance intention (Eckhardt et al., 2009; Im et al., 2011). Additionally, others reported that this variable had no significant influence on behavioral intention (San Martín and Herrero, 2012b). A possible explanation for the absent effect is the context of the study, where facilitating conditions cannot be significant if the risk perceptions of users are higher (San Martín and Herrero, 2012a).

Hypothesis 4 Facilitating conditions are positively related to behavioral intention.

3.5 Gamification impact (GI)

Earlier studies asserted that gamified principles are correlated significantly with the user decision to adopt technology (Lister et al., 2014). Prior studies examined their use in mobile banking services (Baptista and Oliveira, 2017) or with student academic productivity (Juan Carlos Fernández and Aranda, 2017), which indicates that the impact of gamification has a considerable influence on behavioral intention, proved by Baptista and Oliveira (2017). It is verified that there is a correlation between gamification scores and the use of health behavior theory, but only motivation is associated with it (Lister et al., 2014). Moreover, researchers reported that gamification uses mechanics (Richter et al., 2015), rewards, and enjoyment (Nicholson, 2015) components to overtake intrinsic motivation. Obviously, extrinsic rewards such as motivations are perceived by using social capital, self-esteem, and enjoyment in gamification to improve performance (Burke, 2012), create engagement through enforcing power, and increase user loyalty (Sarangi and Shah, 2015).

The author assumes that applying gamified elements in a nongame context, where mHealth is one of them, will make a notable impact on the acceptance of citizens. Thus, gamification impact will positively impact the intention to use mHealth. Hence, we propose:

Hypothesis 5 Gamification is positively related to behavioral intention.

3.6 The mediating effect of personal innovativeness (PI)

The personal innovativeness of information technology, represented in three categories—platform quality, advice quality, and interaction quality—showed a direct effect on the adoption of mHealth by users (Hossain, 2016). This element is derived from combining TAM and TPB (Wu et al., 2011). Another study by Agarwal and Prasad (1998) examined personal innovativeness and whether perceptions play a vital role in the intention to use information technology. It also found that personal innovativeness among users has high-level rates of usage intention on the adoption of online banking services (Lassar et al., 2005) and mobile commerce (Lu, 2014). Because innovative citizens develop positive perceptions more than others in terms of technology features, Agarwal and Prasad (1998) and San Martín and Herrero (2012a) studied innovativeness to influence the relationships between the UTAUT model's factors (PE, EE, and SI) and intention to purchase online.

Given this evidence, it is reasonable to assume that adopting mHealth apps is innovative behavior that is influenced by the user's innovativeness in the domain of information technology. Thus, assuming that intention is the best indicator of consumer behavior (Ajzen, 1991; Venkatesh et al., 2003), the following hypotheses emerge about the mediating influence of personal innovativeness between each of the predictors (PE, EE, SI, and FC) and the predicted variable, which is an intention to use mHealth by consumers in self-monitoring health:

Hypothesis 6a Personal innovativeness mediates the relationship between performance expectancy and behavioral intention.

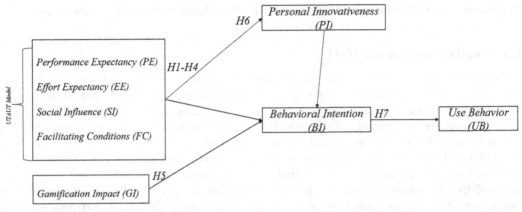

FIG. 1   Research model.

Hypothesis 6b Personal innovativeness mediates the relationship between effort expectancy and behavioral intention.

Hypothesis 6c Personal innovativeness mediates the relationship between social influence and behavioral intention.

Hypothesis 6d Personal innovativeness mediates the relationship between the facilitating conditions and behavioral intention.

3.7 Use behavior (UB)

Venkatesh et al. (2003) affirmed a relationship between behavioral intention and technology use; likewise, Yu (2012) confirmed the same. Hence, the dominant objective of mHealth applications is to make individuals use them. Based on that, the following hypothesis is proposed (Fig. 1):

Hypothesis 7 Behavioral intention is positively related to use behavior.

4. Methodology

4.1 Measures

For collecting data, an online survey was conducted to assess the proposed determinants that influence the success of using mHealth by evaluating the degree of actual use. The adoption diffusion behavior of mHealth is a subjective issue. Therefore, the community of mHealth developers will gain a better understanding of the change in beliefs that consumers established during the use of mHealth. Questionnaire items used in this study to measure the constructs were adapted from previous studies to guarantee the validity of the results. The specific measures in the questionnaire and their sources are shown in Appendix A: Measurement items.

Due to the revisions of those items and to keep consistency with the amended concepts of the proposed determinants of mHealth, the preliminary questionnaire was pretested twice. For the first pretesting, two undergraduates majoring in information technology read and examined each question. They were then asked to revise the questions they found ambiguous or improper. Second, a group of professors from Technological Information College and the College of Business reviewed the survey items. They were asked to provide feedback on the survey questions and to revise the questions further. The final questionnaire consisted of three parts. First, a cover letter was attached to explain the study's purpose and obtain consent from respondents. The second part was associated with the respondent's demographic information, which sought to establish the descriptive characteristics of the sample. Respondents were asked to provide information regarding their age, gender, education, monthly household income, the platform used, and health conditions. The main part, including the items for constructs, was then presented to measure the respondent's perception of each item. The survey was developed in English and all items were measured using a seven-point Likert scale ranging from 1 (strongly disagree) to 7 (strongly agree).

4.2 Sample and data collection process

The survey sample was selected from the population of citizens who use an mHealth app to gather a sufficient sample size for this study. The study required a respondent to have some experience in using mHealth apps for at least 1 month and his/her age to be 18 years or above. A random sampling method was used for this study. Of 268 invitations sent to users through emails and social media, 220 responses were received. All responses were scrutinized, and those containing incomplete responses were discarded. A total of 198 usable responses were obtained. Then, Cronbach's α values were calculated to examine the internal consistency among the items.

5. Analysis

5.1 Data analysis

The analysis was carried out to test and validate the proposed research model using structural equation modeling (SEM). The partial least square (PLS) method is employed to estimate the research model due to its suitability as a statistical analysis technique that has been widely accepted to gauge the validity of theories with empirical data (Götz et al., 2010) and as a tool in information technology fields (Chin et al., 2003). The use of PLS is appropriate for this study specifically because the research model is in its early development phase and not being extensively tested (Teo et al., 2003), the sample size conducted in this research is small, and data are categorical with an unknown nonnormal frequency distribution.

The data analyses were performed using IBM SPSS Amos software (version 22), one of the well-known software applications for PLS-SEM and SPSS (Blunch, 2012). AMOS is a powerful tool that allows the simultaneous analysis of indicator variables (Tabachnick and Fidell, 2013). Data from questionnaires were captured into SPSS (version 20) and imported into Amos software for statistical analysis. As per the PLS procedure, both assessments of the measurement model and structural model have been examined (Hair et al., 2014). The mediating effect of personal innovativeness among PE, EE, SI, and FC with BI has been examined. As implications, the findings of this research will help to understand the factors that influence the use of mHealth apps and provide insights into developing and promoting mHealth apps to maintain health self-management and then reduce chronic diseases.

5.2 Sample characteristics

The sample of this study consists of 198 users, 88 were males (44.4%) and 110 were female (55.6%), as shown in Table 2. The survey respondents' ages were 15.2% below 20 years, 39.4% between 21 and 30 years, 27.3% between 31 and 40 years, 15.2% between 41 and 50 years, and 3.0% 51 years and older. The majority of respondents (33.3%) considered themselves to be below 20,000 average monthly household income (AED), and 28.3% considered themselves to have an average income of 20,000–39,000. In terms of the level of education, high school and diploma holders accounted for 28.3% each while the majority of respondents were undergraduates with 38.4% while 33.3% were postgraduate students. Most users adopted iOS devices (61.6%), and only 38.4% used Android devices. According to the health conditions, almost 70% are in good health and the rest have some health issues such as diabetes, depression, and being overweight.

5.3 Measurement model

The constructs' internal reliability and convergent validity were examined to assess the measurement model (Hair et al., 2014; Leguina, 2015; Ullah et al., 2021) (Table 3). The Cronbach's alpha (α), item-total correlation, factor loadings, composite reliability (CR), and average variance extracted (AVE) obtained in this study are shown in Table 4. The α and CR values were calculated to evaluate the internal reliability at 0.70 (Cronbach et al., 1972). The values have acceptable internal consistency with a range of 0.874–0.927 for α and 0.799–0.929 for CR. Thus, all constructs in the model indicated sufficient reliability confirming the internal reliability of the scores, and these instruments could be relied upon (Hair, 1998).

The method proposed by Fornell and Larcker (1981) was used for convergent and discriminant validity. If the minimum cut-off level is 0.6 for item loading referring (Igbaria et al., 1995) and the values of AVE have at least 0.50 (Hair, 1998), then it can be assumed to present convergent validity. In this study, the estimated loading values ranged from 0.6 to 0.929, the AVE values exceeded 0.5, and both are greater than the recommended levels. Thus, because of the generally good results, the conditions of convergent validity of the measurement model are considered adequate.

In addition, the measurement model was found to have strong internal reliability and convergent validity; the model also showed robust discriminant validity. The satisfactory nature of discriminant validity was assessed as the square root of the AVE of a construct should be greater than its correlations between the factors (Hair et al., 2013). The extracted square root of AVE, which is shown in Table 5, reveals that it was higher than the corresponding correlation between factors, confirming the discriminant validity of the data (Henseler et al., 2009).

TABLE 2 Descriptive statistics of the respondents ($n = 198$).

Demographics	Number	Percentage
Gender		
Male	88	44.4
Female	110	55.6
Age (years)		
Below 20	30	15.2
21–30	78	39.4
31–40	54	27.3
41–50	30	15.2
51 and above	6	3.0
Education		
High school or less	26	13.1
Diploma	30	15.2
Bachelor	76	38.4
Master or higher	66	33.3
Avg. monthly household income (AED)		
Below 20,000	66	33.3
20–000–39,999	56	28.3
40,000–59,999	38	19.2
60,000–79,999	16	8.1
80,000–99,000	12	6.1
100,000 and above	10	5.1
Platform		
iOS (Apple)	122	61.6
Android	76	38.4
Health Conditions		
No health issues	126	63.6

According to Hair et al. (1998), the results confirm both the convergent validity of the factorial model (i.e., the standardized coefficients are significant and above 0.7) and its discriminate validity (i.e., the confidence intervals for the correlations between the latent factor pairs do not include the unit). Consequently, the multiitem scales are reliable and valid for measuring the variables of the theoretical model.

B. Enabling technologies for digital transformation (an example-oriented approach)

TABLE 3 Validity and reliability for constructs.

| Construct | Item | Internal reliability | | Convergent validity | | Average variance extracted |
		Cronbach's alpha	Item-total correlation	Factor loading	Composite reliability	
Performance Expectancy	PE1	0.900	0.789	0.623	0.799	0.503
	PE2		0.873	0.787		
	PE3		0.723	0.590		
	PE4		0.735	0.811		
Effort Expectancy	EE1	0.927	0.808	0.850	0.900	0.694
	EE2		0.871	0.920		
	EE3		0.802	0.778		
	EE4		0.843	0.776		
Social Influence	SI1	0.915	0.833	0.882	0.925	0.804
	SI2		0.825	0.921		
	SI3		0.833	0.888		
Facilitating Conditions	FC1	0.874	0.766	0.831	0.878	0.644
	FC2		0.683	0.733		
	FC3		0.662	0.812		
	FC4		0.820	0.832		
Gamification Impact	GI1	0.919	0.805	0.881	0.929	0.815
	GI2		0.834	0.899		
	GI3		0.871	0.929		
Personal Innovativeness	PI1	0.890	0.692	0.697	0.882	0.653
	PI2		0.763	0.840		
	PI3		0.805	0.855		
	PI4		0.782	0.831		
Behavioral Intention	BI1	0.908	0.813	0.768	0.827	0.546
	BI2		0.826	0.682		
	BI3		0.874	0.784		
	BI4		0.680	0.719		
Use Behavior	UB1	0.905	0.711	0.711	0.811	0.589
	UB2		0.851	0.799		
	UB3		0.890	0.790		

TABLE 4 Correlation matrix.

Construct	Mean	SD	UB	BI	PE	EE	SI	FC	GI
UB	4.327	1.761	**0.767**						
BI	4.975	1.518	0.765	**0.739**					
PE	5.179	1.317	0.437	0.564	**0.709**				
EE	5.152	1.262	0.368	0.451	0.543	**0.833**			
SI	4.381	1.436	0.245	0.251	0.409	0.169	**0.897**		
FC	5.018	1.377	0.350	0.445	0.495	0.246	0.053	**0.802**	
GI	5.071	1.487	0.336	0.338	0.184	0.176	0.062	0.279	**0.903**

Note: Numbers shown in bold indicate the square root of AVE.

TABLE 5 Summary of hypotheses tests.

Hypothesis	Estimate	Significance	Results
H1: PE→BI	0.458	0.006	Supported
H2: EE→BI	0.004	0.973	Not supported
H3: SI→BI	0.210	0.800	Not supported
H4: FC→BI	0.217	0.016	Supported
H5: GI→BI	0.134	0.052	Supported
H6: PI→BI	0.452	0.062	Supported
H6a: PE→PI→BI	−0.017	0.778	Not supported
H6b: EE→PI→BI	0.194	0.008	Supported full mediation
H6c: SI→PI→BI	0.068	0.062	Not Supported
H6d: FC→PI→BI	0.004	0.970	Not Supported
H7: BI→UB	0.707	0.062	Supported

Standardized coefficients; P < .01; P < .05; P < .001.

5.4 Structural model testing

To overcome the individual limitations of the fit measures, a comparison of three measures is recommended to provide a reasonably good estimate of the overall fit (Mehmood et al., 2022a). Generally, indices of NFI and CFI greater than 0.90 (Hair et al., 2010; Mehmood et al., 2022b) and RMSEA less than 0.08 indicate a good model fit. When conducting these indices of the structural model of this study, all model-fitting indices appeared to fulfill the recommended criteria.

For further analysis and to answer the research questions, a structural equation model was developed to identify the relationship between exogenous variables (PE, EE, SI, FC, and GI)

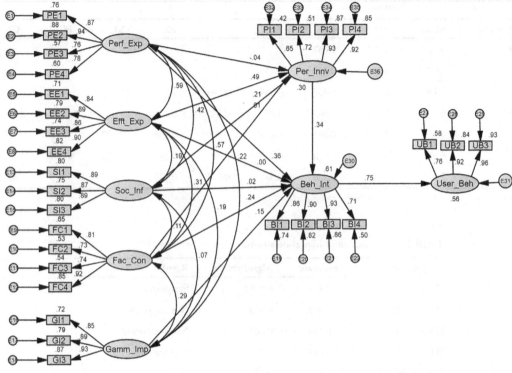

FIG. 2   Structural model.

and endogenous variables (BI and UB). The study tested the relationships among constructs in the hypothesized model by calculating the path coefficient, and the results are displayed in Fig. 2. The results show that the relationships between PE and BI ($\beta=0.36$, $P<.01$), FC and BI ($\beta=0.24$, $P<.01$), GI and BI ($\beta=0.15$, $P<.01$), and BI and UB ($\beta=0.75$, $P<.05$) were significant. Thus H1, H4, H5, and H6 were supported. However, the relationships between EE and BI ($\beta=0.00$, $P>.05$) and SI and BI ($\beta=0.02$, $P>.05$) were insignificant and hence H2 and H3 are not supported in the current study.

The mediating variable PI enters the model to test the effect on the relationships among the determinants of the UTAUT model (including PE, EE, SI, and FC) and BI. In the first stage, it is proven that PI had an influence on BI ($\beta=0.34$, $P<0.05$). The mediating effects of PI were tested among the relationships of PE, EE, SI, and FC on behavioral intention. The model showed that the only significant effect of EE and BI is in the presence of mediator PI so an indirect relationship is available, which means PI has fully mediated the relationship between these two factors because the direct effect of EE and BI is not significant.

6. Discussion

Given the increasing interest in mHealth technology within the field of health self-management, and the need for an overview to guide future research efforts, this research was conducted to determine user behavioral intention to use this technology. This study applied the UTAUT model, gamification impact, and personal innovativeness as a mediator between UTAUT determinants and behavioral intention. The study findings reveal that PE and FC from the UTAUT model were the two significant predictors of the intention to use mHealth apps to improve self-healthcare. Prior studies in various contexts have consistently identified PE as one of the most critical determinants of technology acceptance (Maduku, 2015; San Martín and Herrero, 2012b), highlighting that individuals need to be informed about the advantages of mHealth app adoption to improve health conditions. This implies that for UAE users to adopt mHealth apps, they must first be convinced that the activity would be beneficial to the self-management of their health. Thus, technology developers' advertising initiatives and marketing campaigns may have great relevance in boosting the use of mHealth apps. In contrast, EE and SI did not affect user intention to engage in mHealth apps. The magnitude of the correlation between EE and BI was almost the same as that between FC and BI. These results are inconsistent with previous studies conducted by Kijsanayotin et al. (2009) and Hoque and Sorwar (2017) that found people were encouraged to use health information technology because of its simple design and the influence of social relationships.

It was proven from the analysis that PI has a significant effect on the relationship between the EE construct and the BI. However, PI has no impact of PE, SI, and FC on BI. These results are in line with previous research, especially those that studied the impact of using UTAUT constructs on behavioral intention to use mHealth apps (Agarwal and Prasad, 1998; San Martín and Herrero, 2012a).

This study has reported the benefits of gamification in mHealth apps in an emerging Arab country context and affirms the previous studies that considered goal settings feature, tracking and personalized information, intangible rewards, and reminder features were perceived to be very helpful in behavior change in the health apps and mostly liked by consumers (Anderson et al., 2016; Dennis and O'Toole, 2014; Peng et al., 2016). Therefore, gamification in the technology design plays a critical role in the intention to use it (Rodrigues et al., 2014). Overall, these gamification features need to be developed appropriately to ensure the continued use of this technology.

7. Conclusion

This study found that mHealth apps are commonly adopted and used by various people to improve their self-healthcare. The main factors of the technology acceptance model that impact the actual use of this technology are performance expectancy, effort expectancy, social influence, and facilitating conditions. People have concerns about their privacy and the quality of data provided by mHealth apps, and this is measured by the degree of individual trust. They can also be motivated through the gamified features that are available on the technology by rewards and goal setting. Thus, researchers and technology developers can benefit from these results to provide better service.

Although additional variables for mHealth app adoption were identified, there are still unexplored factors that can be implemented in this context. For instance, intrinsic motivation has a good impact on mHealth app adoption (Kuo and Chuang, 2016). The size of the sample was small to generalize the findings. Also, this study explored only experienced users' behavior toward mHealth apps. Hence, more investigations can provide guidelines to attract new customers and retain existing ones by including nonuser behavior as well. Moreover, researchers may further explore the possible effect of other contextual factors (e.g., cultural values) to uncover other contingencies that may affect the intention of mHealth use and the actual use relationship.

Appendix A: Measurement items

Construct	Corresponding items	Adapted from
Performance Expectancy (PE) (four items)	PE1. I find mHealth app useful in my daily life to enhance my health	Kijsanayotin et al. (2009)
	PE2. Using mHealth app helps to achieve my health goals easier	
	PE3. It is convenient for me to use mHealth app anywhere at any time	
	PE4. It is easy to monitor what I aim to accomplish using mHealth app	
Effort Expectancy (EE) (four items)	EE1. I find it easy to use mHealth app	Carter et al. (2011)
	EE2. Learning how to use mHealth app is easy for me	
	EE3. It is easy for me to become skillful at using mHealth app	
	EE4. My interaction with mHealth app is understandable	
Social Influence (SI) (three items)	SI1. People who influence my behavior influence me to use mHealth app	Venkatesh et al. (2012), Chong et al. (2012), Carter et al. (2011)
	SI2. People who are important to me influence me to use mHealth app	
	SI3. People who are in my social circle influence me to use mHealth app	
Facilitating Conditions (FC) (four items)	FC1. I have the necessary resources to enable me to use mHealth app	Taylor (2015), San Martín and Herrero (2012b)
	FC2. My health environment supports me using mHealth app	
	FC3. Assistance is available when I experience problems with using mHealth app	
	FC4. Using mHealth app is compatible with my life	
	GI1. If mHealth was more fun/enjoyable I would probably use it more often	Baptista and Oliveira (2017)

Construct	Corresponding items	Adapted from
Gamification Impact (GI) (four items)	GI2. If using mHealth gave me points, rewards, and prizes, I would probably use it more often	
	GI3. If mHealth apps were more fun/enjoyable I would probably advise others to use it	
Personal Innovativeness (PI) (four items)	PI1. If I heard about a new mHealth app, I would look for ways to experiment with it	Agarwal and Prasad (1998)
	PI2. Among my peers, I am usually the first to try out new information technology related to mHealth apps	
	PI3. I like to experiment with mHealth app	
	PI4. In general, I am hesitant to try out mHealth app	
Behavioral Intention (BI) (four items)	BI1. I intend to (continue/start) using mHealth app in the future	Escobar-Rodríguez and Carvajal-Trujillo (2013), Yu et al. (2006)
	BI2. I (prefer/would prefer) using mHealth apps to traditional healthcare services	
	BI3. I plan to (continue/start) using mHealth apps frequently	
	BI4. I encourage my colleagues to (start/continue) using mHealth apps	
User Behavior (UB) (three items)	UB1. I am adopting the mHealth app	Dwivedi et al. (2016), San Martín and Herrero (2012a), Baptista and Oliveira (2017)
	UB2. I am using the mHealth app several times a day	
	UB3. I have been using the mHealth app regularly in the last few months	

References

Abushanab, E., Pearson, J.M., 2007. Internet banking in Jordan: the unified theory of acceptance and use of technology (UTAUT) perspective. J. Syst. Inf. Technol. 9 (1), 78–97. https://doi.org/10.1108/13287260710817700.

Agarwal, R., Prasad, J., 1998. A conceptual and operational definition of personal innovativeness in the domain of information technology. Inf. Syst. Res. 9 (2), 204–215. https://doi.org/10.1287/isre.9.2.204.

Ajzen, I., 1991. The theory of planned behavior. Organ. Behav. Hum. Decis. Process. 50 (2), 179–211.

Alaiad, A., Zhou, L., 2014. The determinants of home healthcare robots adoption: an empirical investigation. Int. J. Med. Inform. 83 (11), 825–840. https://doi.org/10.1016/j.ijmedinf.2014.07.003.

Alaiad, A., Zhou, L., Koru, G., 2014. An exploratory study of home healthcare robots adoption applying the UTAUT model. Int. J. Healthc. Inf. Syst. Inform. 9 (4), 44–59. https://doi.org/10.4018/ijhisi.2014100104.

AlAwadhi, S., Morris, A., 2008. The use of the UTAUT model in the adoption of e-government services in Kuwait. In: Paper presented at the proceedings of the 41st annual Hawaii international conference on system sciences (HICSS 2008), 7-10 Jan. 2008.

Al Badi, Alhosani, K.A., Jabeen, F., Stachowicz-Stanusch, A., Shehzad, N., Amann, W., 2022. Challenges of AI adoption in the UAE healthcare. Vision 26 (2), 193–207.

B. Enabling technologies for digital transformation (an example-oriented approach)

Alkatheeri, H.B., Jabeen, F., Mehmood, K., Santoro, G., 2021. Elucidating the effect of information technology capabilities on organizational performance in UAE: a three-wave moderated-mediation model. Int. J. Emerg. Mark. https://doi.org/10.1108/IJOEM-08-2021-1250.

Almuraqab, N.A.S., Jasimuddin, S.M., 2017. Factors that influence end-users' adoption of smart government services in the UAE: a conceptual framework. Electron. J. Inf. Syst. Eval. 20 (1), 11–23.

Anderson, K., Burford, O., Emmerton, L., 2016. Mobile health apps to facilitate self-care: a qualitative study of user experiences. PLoS One 11 (5), e0156164. https://doi.org/10.1371/journal.pone.0156164.

Arman, A.A., Hartati, S., 2015. Development of user acceptance model for Electronic Medical Record system. In: Paper Presented at the International Conference on Information Technology Systems and Innovation, Bandung, Indonesia.

Azadi, M., Yousefi, S., Saen, R.F., Shabanpour, H., Jabeen, F., 2023. Forecasting sustainability of healthcare supply chains using deep learning and network data envelopment analysis. J. Bus. Res. 154, 113357.

Baptista, G., Oliveira, T., 2017. Why so serious? Gamification impact in the acceptance of mobile banking services. Internet Res. 27 (1), 118–139. https://doi.org/10.1108/IntR-10-2015-0295.

Bennani, A.-E., Oumlil, R., 2013. Factors fostering IT acceptance by nurses in Morocco: short paper. In: IEEE 7th International Conference on Research Challenges in Information Science (RCIS). IEEE, pp. 1–6.

Blunch, N., 2012. Introduction to Structural Equation Modelling Using IBM SPSS Statistics and AMOS. Sage Publications.

Burke, B., 2012. Gamification 2020: What Is the Future of Gamification? Gartner, pp. 1–10. Retrieved from Gartner website: www.gartner.com/doc/2226015/gamification-future-gamification.

Burke, B., 2014. Gamify: How Gamification Motivates People to Do Extraordinary Things. Bibliomotion, Inc., Brookline, MA.

Carter, L., Christian Shaupp, L., Hobbs, J., Campbell, R., 2011. The role of security and trust in the adoption of online tax filing. Transform. Gov. People Process Policy 5 (4), 303–318. https://doi.org/10.1108/17506161111173568.

Chang, I.C., Hwang, H.-G., Hung, W.-F., Li, Y.-C., 2007. Physicians' acceptance of pharmacokinetics-based clinical decision support systems. Expert Syst. Appl. 33 (2), 296–303. https://doi.org/10.1016/j.eswa.2006.05.001.

Chatterjee, S., Chaudhuri, R., Vrontis, D., Jabeen, F., 2022. Digital transformation of organization using AI-CRM: from microfoundational perspective with leadership support. J. Bus. Res. 153, 46–58.

Chin, W.W., Marcolin, B.L., Newsted, P.R., 2003. A partial least squares latent variable modeling approach for measuring interaction effects: results from a Monte Carlo simulation study and an electronic-mail emotion/adoption study. Inf. Syst. Res. 14 (2), 189–217. https://doi.org/10.1287/isre.14.2.189.16018.

Chong, A.Y.-L., Chan, F.T.S., Ooi, K.-B., 2012. Predicting consumer decisions to adopt mobile commerce: cross country empirical examination between China and Malaysia. Decis. Support. Syst. 53 (1), 34–43. https://doi.org/10.1016/j.dss.2011.12.001.

Cronbach, L.J., Gleser, G.C., Nanda, H., Rajaratnam, N., 1972. The dependability of behavioral measurements. Theory of generalizability for scores and profiles. Science 178 (4067). https://doi.org/10.1126/science.178.4067.1275. 1275–1275A.

Dajani, D., Yaseen, S.G., 2016. The applicability of technology acceptance models in the Arab business setting. J. Bus. Retail Manag. Res. 10 (3), 46–56.

Davis, F.D., Bagozzi, R.P., Warshaw., 1989. User acceptance of computer technology: a comparison of two theoretical models. Manag. Sci. 35 (8), 982–1003.

Deci, E., Ryan, R., 2004. Handbook of Self-Determination Research. University of Rochester Press, Rochester.

Dehling, T., Gao, F., Schneider, S., Sunyaev, A., 2015. Exploring the far side of mobile health: information security and privacy of mobile health apps on iOS and Android. JMIR mHealth uHealth 3 (1), e3672.

Dennis, T.A., O'Toole, L.J., 2014. Mental health on the go: effects of a gamified attention-bias modification mobile application in trait-anxious adults. Clin. Psychol. Sci. 2 (5), 576–590. https://doi.org/10.1177/2167702614522228.

Deterding, S., Dixon, D., Khaled, R., Nacke, L., 2011. From game design elements to gamefulness: defining "gamification". Proceedings of the 15th International Academic MindTrek Conference: Envisioning Future Media Environments, pp. 9–15.

Devolder, P., Pynoo, B., Sijnave, B., Voet, T., Duyck, P., 2012. Framework for user acceptance: clustering for fine-grained results. Inf. Manag. 49 (5), 233–239. https://doi.org/10.1016/j.im.2012.05.003.

Dwivedi, Y.K., Shareef, M.A., Simintiras, A.C., Lal, B., Weerakkody, V., 2016. A generalised adoption model for services: a cross-country comparison of mobile health (m-health). Gov. Inf. Q. 33 (1), 174–187. https://doi.org/10.1016/j.giq.2015.06.003.

Eckhardt, A., Laumer, S., Weitzel, T., 2009. Who influences whom? Analyzing workplace referents' social influence on IT adoption and non-adoption. J. Inf. Technol. 24 (1), 11–24. https://doi.org/10.1057/jit.2008.31.

Edwards, E.A., Lumsden, J., Rivas, C., Steed, L., Edwards, L.A., Thiyagarajan, A., et al., 2016. Gamification for health promotion: systematic review of behaviour change techniques in smartphone apps. BMJ Open 6 (10), e012447. https://doi.org/10.1136/bmjopen-2016-012447.

Escobar-Rodríguez, T., Carvajal-Trujillo, E., 2013. Online drivers of consumer purchase of website airline tickets. J. Air Transp. Manag. 32, 58–64. https://doi.org/10.1016/j.jairtraman.2013.06.018.

Euehun, L., Semi, H., 2015. Determinants of adoption of mobile health services. Online Inf. Rev. 39 (4), 556–573. https://doi.org/10.1108/OIR-01-2015-0007.

Filsecker, M., Hickey, D.T., 2014. A multilevel analysis of the effects of external rewards on elementary students' motivation, engagement and learning in an educational game. Comput. Educ. 75, 136–148. https://doi.org/10.1016/j.compedu.2014.02.008.

Fogg, B., Adler, R., 2009. Texting 4 Health: A Simple, Powerful Way to Change Lives. Stanford University, Stanford.

Fornell, C., Larcker, D.F., 1981. Evaluating structural equation models with unobservable variables and measurement error. J. Mark. Res. 18 (1), 39–50.

Fox, S., Duggan, M., 2012. Mobile Health. Retrieved from http://www.pewinternet.org/2012/11/08/mobile-health-2012/.

Gao, Y., Li, H., Luo, Y., 2015. An empirical study of wearable technology acceptance in healthcare. Ind. Manag. Data Syst. 115 (9), 1704–1723. https://doi.org/10.1108/IMDS-03-2015-0087.

Gartner, I., 2014. Gartner Says by 2017, Mobile Users Will Provide Personalized Data Streams to More Than 100 Apps and Services Every Day. Retrieved from https://www.gartner.com/newsroom/id/2654115.

Gefen, D., Karahanna, E., Straub, D.W., 2003. Trust and TAMin online shopping: an integrated model. MIS Q. 7 (1), 51–90.

Götz, O., Liehr-Gobbers, K., Krafft, M., 2010. Evaluation of structural equation models using the partial least squares (PLS) approach. In: Esposito Vinzi, V., Chin, W.W., Henseler, J., Wang, H. (Eds.), Handbook of Partial Least Squares: Concepts, Methods and Applications. Springer Berlin Heidelberg, Berlin, Heidelberg, pp. 691–711.

Götze, M., 2015. Increasing User Motivation of a Mobile Health Application Based on Applying Operant Conditioning.

Hair, J.F., 1998. Multivariate Data Analysis, fifth ed. Prentice Hall, Upper Saddle River, NJ.

Hair, J.F., Anderson, R.E., Tatham, R.L., Black, W.C., 1998. Multivariate Data Analysis. Prentice-Hall, Inc, Upper Saddle River, NJ.

Hair, J.F., Anderson, R.E., Babin, B.J., Black, W.C., 2010. Multivariate Data Analysis: A Global Perspective. Pearson, Upper Saddle River, NJ.

Hair Jr., J.F., Hult, G.T.M., Ringle, C., Sarstedt, M., 2013. A Primer on Partial Least Squares Structural Equation Modeling (PLS-SEM).

Hair, J.F., Hult, T., Ringle, C., Sarstedt, M., 2014. A Primer on Partial Least Squares Structural Equation Modeling.

Henseler, J., Ringle, C.M., Sinkovics, R.R., 2009. The Use of Partial Least Squares Path Modeling in International Marketing. vol. 20 Emerald Group Publishing Limited, pp. 277–319.

Hird, N., Ghosh, S., Kitano, H., 2016. Digital health revolution: perfect storm or perfect opportunity for pharmaceutical R&D? Drug Discov. Today 21 (6), 900–911.

Hoque, R., Sorwar, G., 2017. Understanding factors influencing the adoption of mHealth by the elderly: an extension of the UTAUT model. Int. J. Med. Inform. 101, 75–84. https://doi.org/10.1016/j.ijmedinf.2017.02.002.

Hossain, M.A., 2016. Assessing m-Health success in Bangladesh: an empirical investigation using IS success models. J. Enterp. Inf. Manag. 29 (5), 774–796. https://doi.org/10.1108/JEIM-02-2014-0013.

Igbaria, M., Guimaraes, T., Davis, G.B., 1995. Testing the determinants of microcomputer usage via a structural equation model. J. Manag. Inf. Syst. 11 (4), 87–114. https://doi.org/10.1080/07421222.1995.11518061.

Im, I., Hong, S., Kang, M.S., 2011. An international comparison of technology adoption: testing the UTAUT model. Inf. Manag. 48 (1), 1–8. https://doi.org/10.1016/j.im.2010.09.001.

Juan Carlos Fernández, Z., Aranda, D.A., 2017. Implementation of a gamification platform in a master degree (Master in Economics). In: Working papers on Operations Management. vol. 8, pp. 181–190, https://doi.org/10.4995/wpom.v8i0.7431.

Kaba, B., Touré, B., 2014. Understanding information and communication technology behavioral intention to use: applying the UTAUT model to social networking site adoption by young people in a least developed country. J. Assoc. Inf. Sci. Technol. 65 (8), 1662–1674. https://doi.org/10.1002/asi.23069.

B. Enabling technologies for digital transformation (an example-oriented approach)

Kappen, D., Orji, R., 2017. Gamified and persuasive systems as behavior change agents for health and wellness. XRDS: Crossroads, The ACM Magazine for Students 24 (1), 52–55. https://doi.org/10.1145/3123750.

Kijsanayotin, B., Pannarunothai, S., Speedie, S.M., 2009. Factors influencing health information technology adoption in Thailand's community health centers: applying the UTAUT model. Int. J. Med. Inform. 78 (6), 404–416. https://doi.org/10.1016/j.ijmedinf.2008.12.005.

Kim, K.J., Shin, D.-H., 2015. An acceptance model for smart watches: implications for the adoption of future wearable technology. Internet Res. 25 (4), 527–541. https://doi.org/10.1108/IntR-05-2014-0126.

Koivisto, J., Hamari, J., 2014. Demographic differences in perceived benefits from gamification. Comput. Hum. Behav. 35, 179–188. https://doi.org/10.1016/j.chb.2014.03.007.

Koivumäki, T., Ristola, A., Kesti, M., 2008. The perceptions towards mobile services: an empirical analysis of the role of use facilitators. Pers. Ubiquit. Comput. 12 (1), 67–75. https://doi.org/10.1007/s00779-006-0128-x.

Kumar, K.A., Acharjya, B., 2017. Understanding behavioural intention for adoption of mobile games. ASBM J. Manag. 10 (1), 6.

Kuo, M.-S., Chuang, T.-Y., 2016. How gamification motivates visits and engagement for online academic dissemination—an empirical study. Comput. Hum. Behav. 55, 16–27. https://doi.org/10.1016/j.chb.2015.08.025.

Lassar, W.M., Manolis, C., Lassar, S.S., 2005. The relationship between consumer innovativeness, personal characteristics, and online banking adoption. Int. J. Bank Mark. 23 (2), 176–199.

Lee, C.-Y., Tsao, C.-H., Chang, W.-C., 2015. The relationship between attitude toward using and customer satisfaction with mobile application services: an empirical study from the life insurance industry. J. Enterp. Inf. Manag. 28 (5), 680–697. https://doi.org/10.1108/JEIM-07-2014-0077.

Lee, C., Lee, K., Lee, D., 2017. Mobile healthcare applications and gamification for sustained health maintenance. Sustainability 9 (5), 772. https://doi.org/10.3390/su9050772.

Leguina, A., 2015. A Primer on Partial Least Squares Structural Equation Modeling (PLS-SEM). vol. 38 Routledge, pp. 220–221.

Lewis, T., Synowiec, C., Lagomarsino, G., Schweitzer, J., 2012. E-health in low- and middle-income countries: findings from the Center for Health Market Innovations. Bull. World Health Organ. 90 (5), 332.

Lister, C., West, J.H., Cannon, B., Sax, T., Brodegard, D., 2014. Just a fad? Gamification in health and fitness apps. JMIR Serious Games 2 (2), e9. https://doi.org/10.2196/games.3413.

Lu, J., 2014. Are personal innovativeness and social influence critical to continue with mobile commerce? Internet Res.

Maduku, D.K., 2015. An empirical investigation of students' behavioural intention to use e-books. Manag. Dyn. 24 (3), 2.

Mallat, N., 2007. Exploring consumer adoption of mobile payments—a qualitative study. J. Strateg. Inf. Syst. 16 (4), 413–432. https://doi.org/10.1016/j.jsis.2007.08.001.

Anon., 2012. Measuring the Information Society. Place des Nations, Switzerland.

Mehmood, K., Jabeen, F., Iftikhar, Y., Yan, M., Khan, A.N., AlNahyan, M.T., Alkindi, H.A., Alhammadi, B.A., 2022a. Elucidating the effects of organisational practices on innovative work behavior in UAE public sector organisations: the mediating role of employees' wellbeing. Appl. Psychol. Health Well-Being.

Mehmood, K., Jabeen, F., Al Hammadi, K.I.S., Al Hammadi, A., Iftikhar, Y., AlNahyan, M.T., 2022b. Disentangling employees' passion and work-related outcomes through the lens of cross-cultural examination: a two-wave empirical study. Int. J. Manpow. https://doi.org/10.1108/IJM-11-2020-0532.

Mendiola, M.F., Kalnicki, M., Lindenauer, S., 2015. Valuable features in mobile health apps for patients and consumers: content analysis of apps and user ratings. JMIR mHealth uHealth 3 (2), e4283.

Miller, J., 2014. The Future of mHealth Goes Well Beyond Fitness Apps. Retrieved from http://www.cio.com/article/2855047/healthcare/the-future-of-mhealth-goes-wellbeyond-fitness-apps.html.

Mosweu, O., Bwalya, K., Mutshewa, A., 2016. Examining factors affecting the adoption and usage of document workflow management system (DWMS) using the UTAUT model: case of Botswana. Rec. Manag. J. 26 (1), 38–67. https://doi.org/10.1108/RMJ-03-2015-0012.

Nicholson, S., 2015. A RECIPE for meaningful gamification. In: Reiners, T., Wood, L.C. (Eds.), Gamification in Education and Business. Springer, Switzerland, https://doi.org/10.1007/978-3-319-10208-5.

Nieroda, M., Keeling, K., Keeling, D., 2015. Healthcare self-management tools: promotion or prevention regulatory focus? A scale (PR-PV) development and validation. J. Mark. Theory Pract. 23 (1), 57–74. https://doi.org/10.1080/10696679.2015.980174.

Park, J., Yang, S., Lehto, X., 2007. Adoption of mobile technologies for Chinese consumers. J. Electron. Commer. Res. 8 (3), 196.

Park, E., Kim, K.J., Kwon, S.J., 2016. Understanding the emergence of wearable devices as next-generation tools for health communication. Inf. Technol. People 29 (4), 717–732. https://doi.org/10.1108/ITP-04-2015-0096.

Pavlou, P.A., Fygenson, M., 2006. Understanding and predicting electronic commerce adoption: an extension of the theory of planned behavior. MIS Q. 30 (1), 115–143.

Peekhaus, W., 2008. Personal health information in Canada: a comparison of citizen expectations and legislation. Gov. Inf. Q. 25 (4), 669–698. https://doi.org/10.1016/j.giq.2007.05.002.

Peng, W., Kanthawala, S., Yuan, S., Hussain, S.A., 2016. A qualitative study of user perceptions of mobile health apps. BMC Public Health 16 (1), 1–11. https://doi.org/10.1186/s12889-016-3808-0.

Phichitchaisopa, N., Naenna, T., 2013. Factors affecting the adoption of healthcare information technology. EXCLI J. 12, 413–436. https://doi.org/10.17877/DE290R-5602.

Richter, G., Raban, D.R., Rafaeli, S., 2015. Studying gamification: the effect of rewards and incentives on motivation. In: Reiners, T., Wood, L.C. (Eds.), Gamification in Education and Business. Springer, Switzerland.

Rodrigues, L., Costa, C., Oliveira, A., 2014. How gamification can influence the web design and the customer to use the e-banking systems. Proceedings of the International Conference on Information Systems and Design of Communication, pp. 35–44.

San Martín, H., Herrero, A., 2012a. Influence of the user's psychological factors on the online purchase intention in rural tourism: integrating innovativeness to the UTAUT framework. Tour. Manag. 33 (2), 341. https://doi.org/10.1016/j.tourman.2011.04.003.

San Martín, H., Herrero, A., 2012b. Influence of the user's psychological factors on the online purchase intention in rural tourism: integrating innovativeness to the UTAUT framework. Tour. Manag. 33 (2), 341–350.

Sarangi, S., Shah, S., 2015. Individuals, teams and organizations score with gamification: tool can help to motivate employees and boost performance. Hum. Resour. Manag. Int. Dig. 23 (4), 24–27. https://doi.org/10.1108/HRMID-05-2015-0074.

Schaper, L.K., Pervan, G.P., 2006. ICT and OTs: a model of information and communication technology acceptance and utilisation by occupational therapists. Int. J. Med. Inform. 76 (1), S212–S221. https://doi.org/10.1016/j.ijmedinf.2006.05.028.

Scheibe, M., Reichelt, J., Bellmann, M., Kirch, W., 2015. Acceptance factors of mobile apps for diabetes by patients aged 50 or older: a qualitative study. Medicine 2.0 4 (1).

Shareef, M.A., Kumar, V., Kumar, U., Hasin, A.A., 2013. Application of behavioral theory in predicting consumers adoption behavior. J. Inf. Technol. Res. 6 (4), 36–54. https://doi.org/10.4018/jitr.2013100103.

Statista, 2018. Number of mHealth app downloads worldwide From 2013 to 2017 (in billions). In: Medical Technology. Retrieved from https://www.statista.com/statistics/625034/mobile-health-app-downloads/.

Steinhubl, S.R., Muse, E.D., Topol, E.J., 2015. The emerging field of mobile health. Sci. Transl. Med. 7 (283), 283rv3.

Tabachnick, B.G., Fidell, L.S., 2013. Using Multivariate Statistics. Pearson Education, Boston, MA.

Taylor, S.J.E., 2015. Profiling e-health projects in Africa: trends and funding patterns. Inf. Dev. 31 (3), 199–218.

Teo, H.H., Wei, K.K., Benbasat, I., 2003. Predicting intention to adopt Interorganizational linkages: an institutional perspective. MIS Q. 27 (1), 19–49. https://doi.org/10.2307/30036518.

Thirumurthy, H., Lester, R.T., 2012. M-health for health behaviour change in resource-limited settings: applications to HIV care and beyond. Bull. World Health Organ. 90 (5), 390–392. https://doi.org/10.2471/BLT.11.099317.

Ullah, F., Wu, Y., Mehmood, K., Jabeen, F., Iftikhar, Y., Acevedo-Duque, A., Kwan, H.K., 2021. Impact of spectators' perceptions of corporate social responsibility on regional attachment in sports: three-wave indirect effects of spectators' pride and team identification. Sustainability 13 (2), 597.

Venkatesh, V., Morris, M.G., Davis, G.B., Davis, F.D., 2003. User acceptance of information technology: toward a unified view. MIS Q. 27 (3), 425–478.

Venkatesh, V., James, Y.L.T., Xu, X., 2012. Consumer acceptance and use of information technology: extending the unified theory of acceptance and use of technology. MIS Q. 36 (1), 157.

Weerakkody, V., Janssen, M., Dwivedi, Y.K., 2011. Transformational change and business process reengineering (BPR): lessons from the British and Dutch public sector. Gov. Inf. Q. 28 (3), 320–328. https://doi.org/10.1016/j.giq.2010.07.010.

Wen, M.H., 2017. Applying gamification and social network techniques to promote health activities. In: Paper Presented at the 2017 International Conference on Applied System Innovation (ICASI), 13-17 May 2017.

Wu, I.-L., Li, J.-Y., Fu, C.-Y., 2011. The adoption of mobile healthcare by hospital professionals: an integrative perspective. Decis. Support. Syst. 51 (3), 587–596. https://doi.org/10.1016/j.dss.2011.03.003.

B. Enabling technologies for digital transformation (an example-oriented approach)

Yu, C.-S., 2012. Factors affecting individuals to adopt mobile banking: empirical evidence from the UTAUT model. J. Electron. Commer. Res. 13 (2), 104.

Yu, P., Wu, M.X., Yu, H., Xiao, G.Q., 2006. The challenges for the adoption of m-health. In: 2006 IEEE International Conference on Service Operations and Logistics, and Informatics.

Zahra, F., Hussain, A., Mohd, H., 2017. Usability evaluation of mobile applications; where do we stand? AIP Conf. Proc. 1891 (1), 020056.

Zhao, Z., Etemad, S.A., Whitehead, A., Arya, A., 2016. Motivational impacts and sustainability analysis of a wearable-based gamified exercise and fitness system. In: Proceedings of the 2016 Annual Symposium on Computer-Human Interaction in Play Companion Extended Abstracts, 2016.

Further reading

Asimakopoulos, S., Asimakopoulos, G., Spillers, F., 2017. Motivation and user engagement in fitness tracking: heuristics for mobile healthcare wearables. Informatics 4 (1), 5. https://doi.org/10.3390/informatics4010005.

Lim, S., Xue, L., Yen, C.C., Chang, L., Chan, H.C., Tai, B.C., et al., 2011. A study on Singaporean women's acceptance of using mobile phones to seek health information. Int. J. Med. Inform. 80 (12), e189–e202. https://doi.org/10.1016/j.ijmedinf.2011.08.007.

Nunnally, J.C., Bernstein, I.H., 1994. Psychometric Theory. McGraw-Hill, New York.

Nuq, P.A., Aubert, B., 2013. Towards a better understanding of the intention to use ehealth services by medical professionals: the case of developing countries. Int. J. Healthc. Manag. 6 (4), 217–236. https://doi.org/10.1179/2047971913Y.0000000033.

Olff, M., 2015. Mobile mental health: a challenging research agenda. Eur. J. Psychotraumatol. 6 (1), 27882–27888. https://doi.org/10.3402/ejpt.v6.27882.

Zhaohua, D., Shan, L., Oliver, H., 2015. The health information seeking and usage behavior intention of Chinese consumers through mobile phones. Inf. Technol. People 28 (2), 405–423. https://doi.org/10.1108/ITP-03-2014-0053.

Global experiences and approaches to digital transformation in health

Redesigning the global healthcare system through digital transformation: Insights from Saudi Arabia

Nada A. AlTheyab[a], Abdulrahman A. Housawi[a], and Miltiadis D. Lytras[b]

[a]Saudi Commission for Health Specialties, Ministry of Health, Riyadh, Saudi Arabia [b]College of Engineering, Effat University, Jeddah, Saudi Arabia

1. Introduction: Setting the determinants of the health system

According to the World Health Organization (WHO, 2010) "a health system consists of all the organizations, institutions, resources, and people whose primary purpose is to improve health." According to the same report, the building blocks of a modern health system consist of service delivery, health workforce, health information systems, access to essential medicines, financing, and leadership/governance.

The digital transformation of health directly poses significant questions for the added value of digital interventions to the core components of health systems globally. In the next paragraphs, we elaborate with key performance indicators that have been proposed by the WHO and we codesign a context for strategizing the impact of digital transformation in healthcare.

1.1 Health service delivery

The delivery of care and the establishment of delivery services are key pillars of the health system. The detection of inefficiencies and the highest possible optimization of resources must be seen as strategic priorities. With the evolution of digital health, numerous technology-driven challenges such as health service delivery as well as technologies such

TABLE 1 Indicators for healthcare services (WHO, 2010).

Building blocks and indicators

---
1. Health service delivery
   - Number and distribution of health facilities per 10,000 population
   - Number and distribution of inpatient beds per 10,000 population
   - Number of outpatient department visits per 10,000 population
   - General service readiness score for health facilities
   - Proportion of health facilities offering specific services
   - Number and distribution of health facilities offering specific services
   - Specific-services readiness score for health facilities

---

as artificial intelligence (AI), machine learning (ML), the Internet of Things (IoT), robotics, cloud computing, augmented reality, the metaverse, and others create a new digital health services ecosystem. From this point of view, traditional indicators well established for the measurement of health service delivery (see Table 1) should be updated and enhanced.

The critical question is how digital transformation can set up a unified, integrated ecosystem of health services capable of supporting top-quality and value-based healthcare. In this direction, the following objectives must be included in an integrated digital transformation strategy:

- *Dynamic composition of digital health service capacity on demand*: The availability of digital health services integrated with health professionals together with third parties and stakeholders involved in care delivery should be holistically managed within a national context. The capacity of the digitally transformed system to reallocate resources and to reassign responsibilities must be an integral component of the strategy
- *Holistic health data analytics strategy for care delivery*: The development of a robust, comprehensive strategy for healthcare delivery analytics, and the implementation of the required digital services, key performance indicators (KPIs), metrics, and dashboard must lead a new era of patient-centric, value-based healthcare.
- *Standardization and interoperability*: One of the greatest challenges in the current evolution of digital transformation in health systems is the standardization and the interoperability of services. Given the sophisticated context and the complexity of the care providers network within a national context, the sooner the standardization occurs, the fastest the added value of digital transformation will be realized. This is a critical shift in the development of health service delivery on a large scale by standardizing services, processes, and procedures at local levels adopting international best practices and standards.
- *Data governance and unified ecosystem*: While a healthcare information system is another distinct component of the health system, our understanding is that the data governance and the unified ecosystem must be integrated in the health service delivery core function. We need a radical shift on this because data-driven, digitally transformed, value-driven healthcare needs a solid ground on data governance and health services delivery ecosystem (Fig. 1).

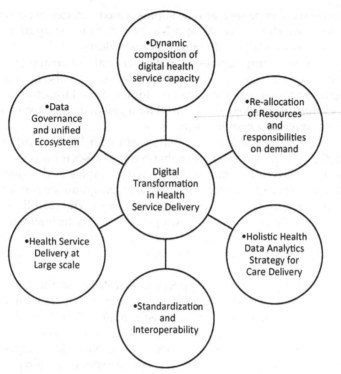

FIG. 1   Digital transformation pillars in health services delivery. *Source: The Authors.*

1.2 Health workforce

The value of the human factor is an inevitable asset for the health system. The integration of a systematic digital transformation strategy for the health workforce has to take into consideration the following well-defined and measurable objectives:

High-quality education: The design, implementation, and delivery of top-quality health education can be enhanced significantly with the exploitation of digital technologies. Establishment of virtual reality and simulation labs, access to digital libraries, and designing technology-enhanced health programs and interventions are a few examples. The idea that the adoption of emerging and streamline technologies can allow the sharing of educational resources and active learning scenarios must be promoted with strategic decisions. Additionally, the design of new training programs, specialties, and roles that will support the key direction of a digitally transformed healthcare is another critical requirement for the new era of health systems. Cloud computing can multidimensionally serve this strategy.

Highly skilled healthcare professionals: The need to integrate highly skilled healthcare professionals in the health system has to exploit novel strategies. Integrating best practices, novel methodologies for technology-enhanced know-how transfer, and skills and competencies building are bold initiatives for digital transformation. The development of agoras of skills and competencies and the establishment of recommendation systems for matching of supply and demand of expertise can be seen as new sophisticated services of competencies management in health systems. The use of AI and ML and can also support new approaches to workflows and services for skills and competencies building.

Certification of skills and competencies is another critical dimension of a health workforce strategy. New smart services can be deployed for this task and also accreditation of training programs and centers will set new requirements for the digital transformation of this aspect. For example, blockchain technology can be used for a global distributed certification network of health workforce skills and competencies.

Professional development is another key component of a health workforce strategy. The establishment of digitally transformed and enhanced processes for professional development will also add value to the entire strategy. Indicative examples in this domain can be smart services for online delivery of focused seminars and programs or services for the transfer of international best practices in professional development. The establishment of academies and health specialties organizations can be supported by new technological capabilities. For example, consider a dynamically composed service of on-demand professional development services per subject. The use of artificial intelligence should also be expected in this high-priority area.

Content and knowledge management supports the strategy for the health workforce. The design and implementation of strategies for knowledge creation, dissemination, and use have to be considered bold aspects of the digital transformation strategy of the health system.

New specialties design in the intersection of health specialties and information technology will be also required. The new era of data-driven decision making supported by the huge capability of data science, AI, and ML will set new requirements and job profiles for the health system.

Incentives for research and development are also another requirement and bold component of the health workforce strategy. Providing digitally supported personal or institutional research and development coworking and coresearching spaces will release significant underutilized resources (Fig. 2).

1.3 Health information systems

The old perception of health information systems (HIS) has now been replaced with the idea of the digital health or eHealth systems. Irrelevant of the used term, it is important to understand that the digital transformation strategy for information systems must take into consideration the following principles:

- Business strategy alignment. The healthcare information systems strategy has to take into serious consideration the overall business strategy of the health system at the national and global levels. This clear coexistence and synergetic effort for the implementation of the business strategy will provide the required connecting components in all the aspects of the digital transformation strategy.
- Service level agreements. The health information systems are simple value-adding complements for the orchestration of the well-defined business objectives and goals. The service level agreements master can provide critical input for the implementation of the

FIG. 2   Digital transformation pillars for health workforce. *Source: The Authors.*

digital transformation strategy by connecting developers, owners, users, and other stakeholders of the digital health ecosystem.

- Innovation capability. One of the most significant aspects of the HIS component is the realization of the technology-driven value. The visioning of robust technological health services at the cutting edge of technologies and value proposition is a key prerequisite for bold implementation plans.
- Management of technologies and systems. The complex nature of the health system also requires sophisticated and advanced management of technologies and systems. The high cost associated with HIS has to be analyzed in terms of return on investment and value realization. Digital transformation in this area can set up the orchestration of services and technologies. The strategic reuse or HIS and allocation of resources within the system is another gain. Also, the design of overarching principles and systems of information systems within the health system can be a significant innovation of the digital transformation strategy.
- Stuffing. One of the most critical parameters for the success of the health system and the HIS component is related to stuffing related to skills and competencies related to databases, networks, AI, ML, cloud computing, and all the other domains. One of the most critical success factors is the need to fill in the gap of computer scientists and information systems experts with the health domain and its specificities.
- Integration

1.4 Access to essential medicines

Concerning the access to essential medicines as a core component of the health system, new technologies and HIS can effectively orchestrate the accomplishment of the following three objectives:

- Reservoirs of essential medicines.
- Medicine discovery with sophisticated new technologies such as AI and ML.
- Codesign and coresearching of new essential medicines.

1.5 Financing

The digital transformation of the financing component of the health system provides unlimited opportunities for digitally transformed services and enhancements. The following is a partial and indicative list of digital transformation interventions:

- Cost effectiveness: With the use of systematic analytics and KPI frameworks, data-driven decision making and control can be efficient. The development of a scalable financial accountability data-driven framework can support diverse and scalable implementations.
- Budgeting: In close relation to the previous aspect, budgeting and sourcing can be effectively implemented with digital health platforms. This will require a systematic integration and coexistence of diverse information systems aiming to manage the financial efficiency of expenditures and investments in the health system.
- Financial performance monitoring with the use of data warehouses and integrated services for descriptive, predictive, and prescriptive analytics and with robust reporting and monitoring capabilities.
- Optimization with the use of AI and ML can be a bold use case for the adoption of a digital transformation strategy for the health system.

1.6 Leadership/governance

The leadership/governance core component of the health system operates as an overarching strategic layer. Thus, the digital transformation should not only promote the implementation of the strategy but also should provide novel, innovative ways for the realization of the added value. The following aspects provide a limited set of possible considerations.

- Accountable healthcare organization.
- Value-based health.
- Patient-centric health.
- Governance as a service.
- Digital transformation strategy.

Fig. 3 provides a high-level framework for the digital transformation of the health system. It is organized around 6 pillars and 31 value-adding carriers.

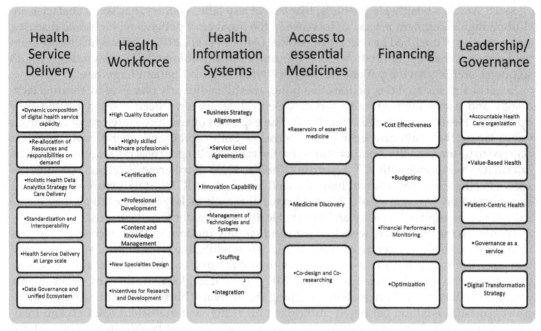

FIG. 3 Digital transformation strategy pillars for health system. *Source: The Authors.*

2. The health system and the digital transformation challenges in Saudi Arabia

The major goal of healthcare facilities is to deliver high-quality medical care to patients. Healthcare providers all around the world strive to please their consumers and achieve the aim of offering this quality of service (Jacobs et al., 2013). The amount of satisfaction with healthcare professionals' services and the fulfillment of their patients' expectations are used to evaluate their performance. Solving the issues the healthcare sector faces has a few specified parts. The quality of healthcare services supplied would be considerably improved if various components, such as organizational structure, were seriously evaluated (Bass and Avolio, 1994). One of the numerous issues confronting the healthcare industry is the rising cost of technology and equipment. Another difficulty is the growing demand for high-quality services from clients (Jacobs et al., 2013). Healthcare providers are finding it increasingly difficult to provide free or low-cost treatments due to these obstacles. Saudi Arabia healthcare companies are now confronting this primary difficulty. In Saudi Arabia, there are three primary organizational entities for healthcare: Ministry of Health (MOH) hospitals, military hospitals, and private hospitals. The MOH is responsible for around 58% of all hospitals in Saudi Arabia. Residents and their relatives are treated for free at these MOH hospitals (Aldakheel and Zakariah, 2021).

Saudi Arabia's healthcare business, like that of any other country, has some obstacles, including high equipment prices, an increase in the number of patients with chronic diseases, and a scarcity of qualified healthcare personnel (Alharbi et al., 2015). Various healthcare companies are embracing information and communication technology (ICT), integrating e-health solutions, and improving patient care, service efficiency, and patient happiness.

Saudi healthcare facilities, on the other hand, are still in the early phases of integrating technology and establishing eHealth services. Several issues have been blamed for the slow uptake of eHealth services (Khalifa, 2013). The expenses of using ICT are one of numerous issues facing the healthcare business. In terms of both capital and operational costs, it is claimed to be more costly. Another reason for the lag in health progress is a scarcity of medical informatics personnel with ICT expertise (Khalifa, 2013). Furthermore, technological issues such as complexity and compatibility as well as a lack of ICT infrastructure obstruct ICT implementation (Alkraiji et al., 2013). Despite these obstacles, Saudi Arabia's healthcare industry is attempting to create and build eHealth services for Saudi citizens.

People, on the other side, wish to create and employ modern medical technology to provide better medical services all over the world. Meanwhile, the costs of achieving this ambition are rising, causing great worry in both public and commercial healthcare sectors throughout the world (Dieleman et al., 2017). By 2023, global healthcare expenses are predicted to have more than doubled, reaching $8.7 trillion. This reflects both rising drug discovery costs ($2.6 billion per newly authorized medicine) and inefficient research and treatment provision (DiMasi et al., 2016). The demand for enhanced healthcare and specialized care programs for the elderly (such as senile dementia and Alzheimer's disease) is growing as people's life expectancies rise across the world (Roser et al., 2019). There are also other issues, such as an increase in the incidence of chronic and lifestyle-related diseases.

The purpose of digital transformation is to alter the way work is done in a unique way that provides outcomes faster and better than traditional techniques while also embracing new ideas and developments to achieve superior conclusions (Kraus et al., 2021). Digital transformation offers effective solutions to establish an efficient, economical, and sustainable society for many stakeholders, such as customers, employees, and beneficiaries, due to the changing style of conducting business. It also affects productivity by altering the working culture. The need for innovation is growing as a result of digital transformation because the need for new goods and services as well as the working environment is continually changing (Gopal et al., 2019). Infrastructure and operational processes as well as marketing are among the adjustments. With the introduction of electronic health records (EHR) for healthcare providers, healthcare is one of many industries benefiting from digital transformation. The main purpose of this implementation is to enhance healthcare by improving patient care by lowering treatment time and concentrating on risk factors by reducing the chance of mistakes as well as improving the entire health system (Belliger and Krieger, 2018).

Healthcare organizations organize and function differently depending on the country; some develop swiftly while others lag. The legal and political policy framework of a country, organizational roles, and human engagement are all elements that influence speed (Haggerty, 2017). The digital transformation of the doctor-patient interaction will also allow new care pathways and significant improvement of the efficiency. ICT technologies for remotely monitoring patient health and the use of Internet of Things applications will also promote new disruptive services of digital health.

2.1 Features of digital transformation

The virtual transformation entails: (1) restructuring the way humans live, work, think, interact, and communicate, primarily based on available technology, with continuous planning and constant seeking of a reformulation of realistic experiences; (2) increasing efficiency, lowering expenses, and rapidly and flexibly introducing new services; and (3) significantly transforming the way humans are supplied with health, education, safety, and security; (4) leveraging modern-day technology to be extra responsive and bendy, with higher predictability and planning for the future; (5) enable quicker innovation to perform preferred and powerful development; and (6) provide a way to provide extra aggressive price, superior teams, and a sustainable lifestyle of innovation (Aldakheel and Zakariah, 2021).

2.2 Digital transformation in Saudi Arabia

The Middle East's difficulties necessitate unprecedented transformation. The reason for this is that the drop in oil prices has prompted the governments of Middle Eastern nations to reconsider and recognize their accomplishments in improving their economies, becoming oil-independent, and effectively adjusting to new circumstances (Alharbi et al., 2015). These nations have experienced significant stress and barriers on multiple occasions, forcing them to swiftly diversify away from their strong reliance on oil and energy earnings. These advancements provide several avenues for diversification, assist improvements in various sectors, and have significant progress targets. Saudi Arabia, for example, established Vision 2030 in 2016. In the near future, these modifications and planned initiatives will undergo a large and ambitious transition, with daring decisions to give greater service to the community (Alkraiji et al., 2013). Vision 2030's primary purpose is to conduct a large-scale modernization program that promotes technical growth and economic variety.

Digital transformation is a critical component in accomplishing this lofty aim (Deloitte, 2018). Smartphones and other technology are increasingly being used, altering established operating paradigms. . Saudi Arabians have quickly accepted smartphone devices and innovations, and the country is now one of the world's largest recipients of digital technology and a leader in the pace of adoption. Because of these quick developments, government agencies will need to upgrade their applications and technology to match the needs of their consumers and members, ushering in a new digital revolution (Jacobs et al., 2013). Companies that do not update themselves risk failing to fulfill customer expectations and failing to manage their businesses efficiently. Because of these consequences, businesses must accept and implement digital transformation. Government agencies must utilize digital technology to suit the needs of the public and the expectations of patients in the healthcare business (Dieleman et al., 2017). The bulk of Saudi Arabian organizations, however, is falling behind in formulating genuine strategies and plans to construct a digital economy. Furthermore, these developments provide a dilemma for many organizations and businesses that are concerned that digital technology may render their activities outdated in future years. According to Altuwaijri (2010), just 4% of commercial companies in Saudi Arabia and the United Arab Emirates are digital leaders, 44% are hesitant to adopt new technologies, and 44% are concerned about how they will affect the future. Some businesses have been outraged by the development of the IoT. Altuwaijri stressed the importance of a national eHealth program in Saudi Arabia and urged the establishment of a national eHealth workplace. Similarly, Alsanea (2012) stressed the significance of EHR at the King Faisal Specialist Hospital and Research Center, and with EHR deployment, organizations performed data analysis on public health benefits, national authorities, and others.

3. Healthcare digital transformation in Saudi Arabia

The Saudi Data and Artificial Intelligence Authority (SDAIA) has teamed with Royal Philips and IQVIA to assist the country's goal of being a leader in AI intelligence in healthcare (Al-Ruithe et al., 2018). The collaborations seek to provide top health technology and analytics tools to the country's data scientists and healthcare professionals to assist them in developing integrated solutions that increase the performance and productivity of healthcare systems. These collaborations get the healthcare industry closer to meeting the Saudi Vision 2030 goals, allowing them to provide even more sophisticated healthcare to the people.

eHealth services have radically altered the way healthcare systems are managed across the world. Health systems have profited by automating many previously manual operations, transitioning from paper to paperless services, and digitizing records so that they are now integrated into a more efficient and standard system. The majority of the population in Saudi Arabia is served by five primary health authorities, with the Ministry of Health overseeing 60% of hospitals, four other agencies overseeing 20%, and the public sector overseeing the other 20% (Al-Hanawi et al., 2018). Because some hospitals have computerized systems and others do not, the MOH, which oversees the majority of healthcare institutions, has been working on integrating hospitals through a national strategy for an eHealth service. Even among hospitals with electronic HIS, the systems are from many manufacturers and are not linked, posing a complex set of hurdles for building a statewide eHealth system (Alsulame et al., 2016).

Healthcare improvement is a key component of Vision 2030, which calls for bettering access to care, raising the value of treatment provided, and strengthening health prevention efforts (Noor, 2019). Saudi Arabia presently has a national healthcare system, with the government providing 78.9% of total healthcare services. To overcome the previously delayed transition to IT and improve the quality of healthcare services, this approach comprises a budget of SR 4 billion (US $1.1 billion). Although EHRs are already used in 40% of MOH hospitals, primary healthcare facilities still utilize paper-based records and are in the process of converting to information technology (Noor, 2019). As a result, the adoption of eHealth in MOH facilities has been sluggish, with some separate information systems running in regional directorates and central hospitals that are not linked to one another (Alsulame et al., 2016). However, in recent years, eHealth has gained popularity in Saudi Arabia, with Makkah City hospitals reporting a large increase in eHealth usage.

eHealth is also an established approach at King Abdulaziz Medical City (KAMC), with eHealth programs being used by 98% of its departments (Zaman et al., 2018). Furthermore, King Faisal Hospital (KFH) has installed an eHealth system in nearly all of its departments while Al Noor General Hospital's eHealth system has 76.9% acceptability as it installs further eHealth applications (Zaman et al., 2018). Because this is insufficient, the MOH is working with the National Transformation Program (NTP) to develop alternative technology funding sources by increasing private sector ownership and utilizing private sector management to improve the efficiency and effectiveness of HIT and digital transformation before Vision 2030 is completed (Zaman et al., 2018).

The MOH has also made initiatives to improve healthcare governance by developing a digital accountability framework (Alharbi, 2018). Furthermore, the MOH's commitment of 4 billion Saudi Riyals (US $1.1 billion) between 2008 and 2011 to develop holistic eHealth programs and improve the maturity status of the country's health information infrastructure provides an economic boost due to the use of technology in the health sector (Alkraiji, 2012).

4. Key challenges and conclusions

According to Zaman et al. (2018), the key hurdles were implementing the eHealth system, staff training and skills, views of the eHealth system, and so on. The authors discovered that employees had difficulty implementing and accepting eHealth systems. This is due to a lack of training and oversight in these areas. Some studies have also found that the technical and computer skills of hospital employees influence the extent of eHealth deployment and use (Alsulame et al., 2016). The technical skills and competence of healthcare workers are critical concerns to address before the implementation of any hospital information system (Hasanain and Cooper, 2014). The most significant social hurdles are a lack of computer proficiency and an aversion to using the new system. Another cited hurdle to electronic health record (EHR) deployment in Saudi Arabia is language. The reason for this is because Arabic is the first language spoken here, while the mechanisms in place are primarily in English.

Resistance to the new technology was one of the two social obstacles identified as hindering EHR adoption in Saudi Arabia (Alanazy, 2006). As part of any organizational change, certain people will reject new systems. Some staff may be hesitant to use new technology like an EHR system. One of the barriers to EHR adoption in various Saudi institutions is this negative attitude (Alkraiji et al., 2013). The lack of computer competence and experience among healthcare staff was the second identified social obstacle (Alanazy, 2006). It was revealed that there is a relationship between the absence of essential understanding of EHR software and computer competence among some physicians and other healthcare practitioners and the required level for EHR adoption success in Saudi Arabia (Alanazy, 2006). Various studies have shown that firms struggle to adopt new systems when personnel lack the requisite skills and knowledge to handle computers and electronic systems, such as EHRs (Hasanain and Cooper, 2014).

Several challenges hamper the implementation of eHealth in Saudi Arabia. Cultural, regulatory, and personnel issues are the biggest barriers to eHealth adoption in Saudi Arabia (Alsulame et al., 2015). For example, the eHealth project was delayed because top management clearance to engage more health information professionals was difficult to secure (Househ and Al-Tuwaijri, 2011). The procurement procedure in Saudi medical institutions, according to Alsulame et al. (2015), has become a hurdle for some Saudi medical practitioners. Several challenges linked to procurement concerns and their influence on hospital eHealth efforts have been found in previous research (Cresswell et al., 2013).

Some organizations were still having trouble raising funds for the eHealth effort. In another investigation, the computerized provider order entry (CPOE) and clinical decision support system (CDSS) were found to be excessively high (Almutairi et al., 2012). The findings show that the National Strategic Plan's implementation roadmap has not been conveyed to healthcare organizations as part of the plan. This is similar to research that looked at the lack of an eHealth strategy and discovered that each firm had its strategy (Al Shorbaji, 2008). This challenge will be resolved after the Saudi Arabian national program is completed. Outsourcing initiatives have been proven to have a detrimental influence on creating team communication within the firm when it comes to coordination. According to Househ and Al-Tuwaijri (2011), various professionals may not be in communication with one another or be aware of the state of the eHealth project owing to inadequate management.

References

Al Shorbaji, N., 2008. E-health in the Eastern Mediterranean Region: a decade of challenges and achievements. East Mediterr. Health J. 14 (Supp), S157–S173.

Alanazy, S., 2006. Factors Associated with the Implementation of Electronic Health Records in Saudi Arabia. University of Medicine and Dentistry of New Jersey.

Aldakheel, E.A., Zakariah, M., 2021. Digital transformation challenges in healthcare and its effect on the patient-physician relationship. J. Manag. Inf. Decis. Sci. 24, 1–11.

Al-Hanawi, M.K., Alsharqi, O., Almazrou, S., Vaidya, K., 2018. Healthcare finance in the Kingdom of Saudi Arabia: a qualitative study of householders' attitudes. Appl. Health Econ. Health Policy 16 (1), 55–64.

Alharbi, M.F., 2018. An analysis of the Saudi healthcare system's readiness to change in the context of the Saudi National Health-care Plan in vision 2030. Int. J. Health Sci. 12 (3), 83.

Alharbi, F., Atkins, A., Stanier, C., 2015, February. Strategic framework for cloud computing decision-making in the healthcare sector in Saudi Arabia. In: The Seventh International Conference on eHealth, Telemedicine, and Social Medicine. Vol. 1, pp. 138–144.

Alkraiji, A., 2012. Issues of the Adoption of HIT-Related Standards at the Decision-Making Stage of Six Tertiary Healthcare Organizations in Saudi Arabia (Doctoral dissertation). © Abdullah Ibrahim Alkraiji.

Alkraiji, A., Jackson, T., Murray, I., 2013. Barriers to the widespread adoption of health data standards: an exploratory qualitative study in tertiary healthcare organizations in Saudi Arabia. J. Med. Syst. 37 (2), 1–13.

Almutairi, M.S., Alseghayyir, R.M., Al-Alshikh, A.A., Arafah, H.M., Househ, M.S., 2012. Implementation of computerized physician order entry (CPOE) with clinical decision support (CDS) features in Riyadh hospitals to improve quality of information. In: Quality of Life through Quality of Information. IOS Press, pp. 776–780.

Al-Ruithe, M., Benkhelifa, E., Hameed, K., 2018. Key issues for embracing the cloud computing to adopt a digital transformation: a study of Saudi public sector. Proc. Comp. Sci. 130, 1037–1043.

Alsanea, N., 2012. The future of health care delivery and the experience of a tertiary care center in Saudi Arabia. Ann. Saudi Med. 32 (2), 117–120.

Alsulame, K., Khalifa, M., Househ, M.S., 2015. eHealth in Saudi Arabia: current trends, challenges and recommendations. ICIMTH 213, 233–236.

Alsulame, K., Khalifa, M., Househ, M., 2016. E-health status in Saudi Arabia: a review of current literature. Health Policy Technol. 5 (2), 204–210.

Altuwaijri, M.M. (2010). Supporting the Saudi e-health initiative: the master of health informatics program at KSAU-HS. East Mediterr. Health J., 16 (1), 119-124, 2010.

Bass, B.M., Avolio, B.J., 1994. Transformational leadership and organizational culture. Int. J. Public Admin. 17 (3–4), 541–554.

Belliger, A., Krieger, D.J., 2018. The digital transformation of healthcare. In: Knowledge Management in Digital Change. Springer, Cham, pp. 311–326.

Cresswell, K., Coleman, J., Slee, A., Williams, R., Sheikh, A., ePrescribing Programme Team, 2013. Investigating and learning lessons from early experiences of implementing ePrescribing systems into NHS hospitals: a questionnaire study. PLoS One 8 (1), e53369.

Deloitte, 2018. The 2018 Global Health Care Outlook: The Evolution of Smart Health Care. Deloitte.

Dieleman, J.L., Campbell, M., Chapin, A., Eldrenkamp, E., Fan, V.Y., Haakenstad, A., Murray, C.J., 2017. Future and potential spending on health 2015–40: development assistance for health, and government, prepaid private, and out-of-pocket health spending in 184 countries. Lancet 389 (10083), 2005–2030.

DiMasi, J.A., Grabowski, H.G., Hansen, R.W., 2016. Innovation in the pharmaceutical industry: new estimates of R&D costs. J. Health Econ. 47, 20–33.

Gopal, G., Suter-Crazzolara, C., Toldo, L., Eberhardt, W., 2019. Digital transformation in healthcare–architectures of present and future information technologies. Clin. Chem. Lab. Med. 57 (3), 328–335.

Haggerty, E., 2017. Healthcare and digital transformation. Netw. Secur. 2017 (8), 7–11.

Hasanain, R.A., Cooper, H., 2014. Solutions to overcome technical and social barriers to electronic health records implementation in Saudi public and private hospitals. J. Health Inform. Dev. Ctries. 8 (1).

Househ, M.S., Al-Tuwaijri, M., 2011, January. Early development of an enterprise health data warehouse. In: ITCH, pp. 122–126.

Jacobs, R., Mannion, R., Davies, H.T., Harrison, S., Konteh, F., Walshe, K., 2013. The relationship between organizational culture and performance in acute hospitals. Soc. Sci. Med. 76, 115–125.

Khalifa, M., 2013. Barriers to health information systems and electronic medical records implementation. A field study of Saudi Arabian hospitals. Proc. Comp. Sci. 21, 335–342.

Kraus, S., Schiavone, F., Pluzhnikova, A., Invernizzi, A.C., 2021. Digital transformation in healthcare: analyzing the current state of research. J. Bus. Res. 123, 557–567.

Noor, A., 2019. The utilization of e-health in the Kingdom of Saudi Arabia. Int. Res. J. Eng. Technol. 6 (09), 11.

Roser, M., Ortiz-Ospina, E., Ritchie, H., 2019. Life Expectancy. Our World in Data. Google Scholar.

World Health Organization, 2010. Monitoring the Building Blocks of Health Systems: A Handbook of Indicators and Their Measurement Strategies. World Health Organization. https://apps.who.int/iris/handle/10665/258734.

Zaman, T.U., Raheem, T.M.A., Alharbi, G.M., Shodri, M.F., Kutbi, A.H., Alotaibi, S.M., Aldaadi, K.S., 2018. E-health and its transformation of the healthcare delivery system in Makkah, Saudi Arabia. Int. J. Med. Res. Health Sci. 7 (5), 76–82.

A strategic framework for digital transformation in healthcare: Insights from the Saudi Commission for Health Specialties

Abdulrahman A. Housawi[a] and Miltiadis D. Lytras[b]

[a]Saudi Commission for Health Specialties, Ministry of Health, Riyadh, Saudi Arabia [b]College of Engineering, Effat University, Jeddah, Saudi Arabia

1. Introduction

Digital transformation appears to be one of the top priorities in innovation and sustainability agendas in the healthcare industry. As a theoretical domain, it brings together areas and ideas related to business performance, efficiency, value-based systems, and technology-driven innovation. The multidisciplinary nature of the phenomenon challenges any effort to justify strategic consultation for its implementation in modern organization. In our work, we approach digital transformation within the context of Vision 2030 in Saudi Arabia, and we interpret key determinants of the international benchmarks for the phenomenon within the strategic priorities of the Saudi Commission for Health Specialties (SCFHS).

Vision 2030 and the digital transformation vision in Saudi Arabia provide a unique context for the translation of the literature and the justification of its impact in the strategic priorities of the SCFHS.

1.1 The profile of the SCFHS

The SCFHS was established in 1992 by Royal Order M/2 to set standards for health practices. Its headquarters is in Riyadh, and several branches are distributed across the country.

The SCFHS vision is "a healthy community through competent healthcare providers based on best standards by protecting and promoting health led by compassionate and competent practitioners."

SCFHS is a versatile professional organization tasked with three main roles related to health professions: (1) regulation and supervision of health professional training, (2) professional accreditation of health professionals, and the regulation of their professional development, and (3) oversight of scientific health councils in Saudi Arabia.

Within its portfolio, SCFHS sets the standards and controls for training health practitioners and provides global oversight of these training activities. The portfolio includes accrediting health professional practices in the Saudi Arabia through classification and registration as well as supervising health professional residency training programs and participating training institutions.

SCFHS is engaged in several key initiatives to deliver on its mission. It is governed by a board of trustees and utilizes several councils and committees of respected and qualified professionals for the services provided.

SCFHS aims to regulate the health care industry, improve professional performance, develop and encourage skills, and enrich scientific theory and practice in different health-related fields by:

1. Developing, approving, and supervising professional health-related programs, and developing sustained medical education programs for health-related disciplines in line with the general educational framework.
2. Forming, supervising, and approving the recommendation of health-scientific boards and subcommittees needed to carry out the tasks assigned to SCFHS.
3. Supervising and approving the results of specialized examinations through specialist scientific committees and boards.
4. Issuing professional certificates, such as diplomas, fellowships, and membership certificates, regardless of whether SCFHS actually administers the examinations or merely contributes to them.
5. Coordinating with other professional health boards, commissions, associations and schools inside and outside the country.
6. Developing the principles and standards governing the practice of health professions, including a code of ethics.
7. Evaluating and approving health certificates.
8. Encouraging health-related scientific research, publishing scientific papers, and issuing its own journals or periodicals in addition to proposing research topics as well as providing support and full or partial funding.
9. Participating in recommending general plans to qualify and develop human resources in health fields.
10. Approving the establishment of scientific associations for health specialties.

In the next section, we elaborate on the literature review and we synthesize complementary aspects of the phenomenon in an effort to integrate these into a strategic methodological framework to study the digital transformation in healthcare.

2. Literature review on digital transformation

The relevant literature of the domain covers diverse aspects of the digital transformation phenomenon (Agrawal, 2020; Bousquet et al., 2021; Dugstad et al., 2019; Haggerty, 2017; Lytras and Visvizi, 2018; Lytras et al., 2017, 2019a,b; Merolli, 2021; Seoane et al., 2021; Steinhauser, 2021; Walsh and Rumsfeld, 2017; Yamamoto, 2021; Yap et al., 2021; Yip et al., 2019; Bayram et al., 2020). The compact literature review in this section is interpreted and exploited in the context of the core function of the SCFHS. A summary of the studied literature is provided in Tables 1 and 2 below.

A well-defined literature stream is related to business process modeling and the integration of people, processes, and policies as a bold approach to the specification of the digital transformation strategy.

In a recent article, Meske et al. (2019) elaborated on *"The potential role of digital nudging in the digital transformation of healthcare."* They introduced the concept of digital nudging that aims to elicit behavior that is beneficial for the individual while at the same time respecting the individual's own preferences and freedom of choice. Digital transformation requires human involvement and the use of sophisticated sociotechnical systems and advanced tools, and thus the development of strategies and policies to help humans and health specialists to deploy procedures and behaviors that promote the added value of technology deployment. In the quest of any modern healthcare organization to translate the strategy of digital transformation into action plans, it is necessary to understand the core business functions as well as the disruptive character of the technology toward introducing innovations and disruptive value-adding services.

TABLE 1   Key aspects of the literature on digital transformation.

Author	Key perspective/focus of the study	Challenge for digital transformation research in SCFHS
Meske et al. (2019)	– Digital nudging – Value creation	In the context of SCFHS, it is significant to deliver research that will inform us how experts from the six core functions interpret technology and how this technology can be beneficial for the delivery of strategic objectives of organizations
Hermann et al. (2018)	– Digital innovation – Value networks – Business models – Technological enablers – Notable transformation of well-established services	In our research, this finding drives our rationale and research aim. The systematic mapping and understanding of the current core functions of the SCFHS and the well-established services serve as the basis for notable transformation based on value delivery and justification

Continued

TABLE 1   Key aspects of the literature on digital transformation—cont'd

Author	Key perspective/focus of the study	Challenge for digital transformation research in SCFHS
Benjamin and Potts (2018)	– Lessons learned from digital transformation in government – Metaphors in healthcare – Achieve sustainable change at a scale	We intend to investigate how the introduction of digital transformation in the core functions of the SCFHS can support sustainable change that will lead to enhanced performance and quality and utilization of costs
Jahankhani and Kendzierskyj (2019)	– Blockchain as an enabler of advanced data privacy and protection	In our study, we want to investigate how technologies like blockchain enrich the value proposition of the SCFCS
Hasan et al. (2020)	– Mobile technology and applications – Design approach – Proof of concept	In our research, we focus on the detailed specification of use cases for value realization of the adoption of disruptive technologies in the six core functions of the SCFHS. For example, how artificial intelligence can enhance the value proposition of key services in training, accreditation, and professional development. Also, how mobile technologies and big data analytics together with cloud computing can enhance the professional development and the workforce planning functions in SCFHS
Woodside and Amiri (2018)	– Value chain in healthcare – Understanding the shift from the traditional healthcare value chain to the Industry 4.0 digital healthcare value chain model	One of the critical priorities in our research study is to identify the current and possible technology-driven enhancement of the value chain in the SCFHS
Hermes et al. (2020)	– Highlight the significance of digital platforms that leverage platform-mediated ecosystems for value cocreation with expanded networks of stakeholders, intermediaries, and providers – The results indicate eight new roles within healthcare: information platforms, data collection technology, market intermediaries, services for remote and on-demand healthcare, augmented and virtual reality providers, blockchain-based PHR, cloud service providers, and intelligent data analysis for healthcare providers – The results further illustrate how these roles transform value proposition, value capture, and value delivery in the healthcare industry	In an organization like SCFHS, understanding a platform-mediated ecosystem of value cocreation in line with the six core functions is a critical success factor for the implementation of the digital transformation strategy

TABLE 2 An overview of the selected literature review and its implication to our research design.

Authors	Title of article	Key contribution
Meske et al. (2019)	The potential role of digital nudging in the digital transformation of the healthcare industry	**Focus**: On digital nudging and the possibilities of this concept in hospitals
Müller (2021)	The digital transformation of healthcare	**Focus:** • On provider and consumer prospects for the use of telemedicine in current and future patient care • On identifying emerging (ethical) conflicts
Meskó et al. (2017)	Digital health is a cultural transformation of traditional healthcare	**Focus:** On digital health **Analysis:** How it affects the status quo of care and the design of the study for the application of technological innovations in the practice of medicine
Herrmann et al. (2018)	Digital transformation and disruption of healthcare sector: Internet-based observational study	**Focus:** On digital and disruptive innovation **Analysis:** What type of technological enablers, business models, and value networks seem to be emerging from different groups of innovators with respect to their digital transformational efforts
Benjamin and Potts (2018)	Digital transformation in government: lessons for digital health?	**Focus:** On describing some common trends in digital transformation in government **Analysis**: how they apply to the health sector, using NHS England as a leading exemplar
Fernández et al. (2020)	Starting the path of digital yransformation in health innovation in digital health.	**Focus:** On information and communication technologies (ICTs) in healthcare **Analysis:** Ensure that information related to healthcare is accessible by the right person, in the right place, at the right time, and in a safe manner, seeking to optimize the efficiency, equity, access, safety, and quality of health care
Jahankhani and Kendzierskyj (2019)	Emerging E-Health MSPs in China, disruptive innovation through digital transformation	**Focus:** On impact of digital transformation on Chinese healthcare industry **Analysis:** Blockchain and its usefulness as a mechanism for unchanged, monitoring, security, and data privacy protection in healthcare transformation

Continued

TABLE 2   An overview of the selected literature review and its implication to our research design—cont'd

Authors	Title of article	Key contribution
Natsis et al. (2020)	Digital transformation of the health sector in Greece: Evaluation of the websites of health organizations in the region of Attica	**Focus:** On websites of all public hospitals and private clinics in the region of Attica **Analysis:** Possible problems and content omissions that may be present in order to be amended or improved
Hasan et al. (2020)	Monile health app for university students with mental illness: a digital transformation design approach from Australia	**Focus:** On introducing a digital transformation approach to design a prototype mobile health application that transforms the current psychological support services available at a university into a mobile digital platform
Woodside and Amiri (2018)	Healthcare hyperchain: Digital transformation in the healthcare value chain	**Focus:** On healthcare value chain in the context of digitally enabled technologies **Analysis:** How hyperledger and blockchain have the potential to extensively alter healthcare organizations
Hermes et al. (2020)	The digital transformation of the healthcare industry: Exploring the rise of emerging platform ecosystems and their influence on the role of patients	**Focus:** On the digital transformation of the healthcare industry. **Analysis:** A general value ecosystem of the digital healthcare industry and validated findings with industry experts from traditional and emerging healthcare sectors
Han et al. (2020)	Emerging E-Health MSPs in China, disruptive innovation through digital transformation	**Focus:** On Six e-health platforms (MSPs) **Analysis:** Uncovers that all parties in the current healthcare ecosystem have already been impacted to some extent by the complementary functions that the emerging MSPs bring about
Peek et al. (2020)	Digital health and care in pandemic times: Impact of COVID-19	**Focus:** On developments in service delivery, artificial intelligence (AI), and data sharing instilled by the COVID-19 crisis
Klinker et al. (2020)	Digital transformation in healthcare: augmented reality for hands-free service innovation, information systems frontiers	**Focus:** On augmented reality smart glass applications **Analysis:** How smart glass prototype supports healthcare professionals during wound treatment by allowing them to document procedures hands-free while they perform them
Warraich et al. (2018)	The digital transformation of medicine can revitalize the patient-clinician relationship	**Focus:** On digital technologies and how they could revitalize the patient-clinician relationship and perhaps also improve clinician well-being
Frennert (2021)	Hitting a moving target: digital transformation and welfare technology in Swedish municipal elder care	**Focus:** On digital transformation and welfare technology in municipal elder care

C. Global experiences and approaches to digital transformation in health

In the context of SCFHS, it is significant to deliver research on how experts from the six core functions interpret technology and how this technology can be beneficial for the delivery of strategic objectives of organizations. According to Meske et al. (2019), *"digital nudging can positively influence the use of technology, new value creation, the change of structures, and consequently (the) financial dimensions of digital transformation, supporting not only caregivers but also caretakers."*

Meskó et al. (2017) brought into the discussion of digital transformation the theme of equal level partnerships between professionals and patients and how this is facilitated by the disruptive technologies of our times. This direction together with their focus on understanding new designs for the utilization of innovative digital solutions also drives our research. With the six core functions of SCFHS, we are interested in researching how partnerships supported by technological systems can drive digital transformation and its relevant value proposition. In our research, we are interested in understanding the key determinants of the disruption in the core functions of the SCFHS as a reflection of the challenges of our times. In this direction, the recent study from Herrmann et al. (2018) analyzed the concept of digital innovation, commenting that healthcare only recently focused on its bold digital transformation, promoting drastic implications and a value-based approach. Their key contribution is the systematic analysis of technological enablers, value networks, and business models, as these are deployed by innovators in digital transformation. They concluded that the incremental enhancement of business processes toward more effective services and notable transformations of well-established services is a straightforward approach to digital transformation while some start-up companies develop new markets and channels. In our research, this finding drives our rationale and research aim. The systematic mapping and understanding of the current core functions of the SCFHS and the well-established services serve as the basis for notable transformations based on value delivery and justification. This is a significant objective in our research.

The literature survey also offers lessons learned from digital government projects in the context of digital health. Benjamin and Potts (2018) emphasized the need to achieve sustainable change at a scale and on the necessity of organizations to approach the challenges faced in digital transformation projects. In our research, in the context of the SCFHs we intend to investigate how the introduction of digital transformation in the core functions of the SCFHS can support sustainable change that will lead to enhanced performance and quality and utilization of the costs.

Jahankhani and Kendzierskyj (2019) discussed the contribution of blockchain in the realization of digital transformation in healthcare. They focus on the significance of blockchain for data protection and privacy and indicate how this integrative strategy can support digital transformation scenarios. In our research, a key priority is investigating how technologies such as blockchain can support value-adding scenarios and services that enrich the value proposition of the SCFHS. Blockchain is only one technology that serves as a focus of our research study. Mobile technologies and applications are other examples. In their study, Hasan et al. (2020) considered mobile health apps as a test bed for digital transformation design approaches.

Their research emphasizes how prototypes can support the proof of concept for the value proposition of digital transformation. In our research, we focus on the detailed specification of use cases for the value realization of the adoption of disruptive technologies in the six core functions of the SCFHS. For example, how artificial intelligence can enhance the value proposition of key services in training, accreditation, and professional development. Also, how mobile technologies and big data analytics together with cloud computing can enhance the professional development and workforce planning functions in SCFHS.

All the previous theoretical contributions and research are related to an integral value chain of digital transformation. This is another critical priority for our research approach. We are interested in defining the value chain of the digital transformation in SCFHS. In this direction, the research work of Woodside and Amiri (2018) is significant. According to their key proposition, all organizations working toward digital transformation need to *reenvision their value chain to improve patient outcomes and be flexible and adaptable in their ability to respond to the competitive marketplace.* Their emphasis is also on digitally enabled technologies such as hyperledger and blockchain. They elaborate on how healthcare organizations can move from the traditional value chain to the Industry 4.0 digital healthcare value chain model. The interesting aspect of this research is that sets the design and update of traditional value chains with digital enablers and outcomes as a priority toward the success of digital transformation in healthcare. From this point of view, one of the critical priorities in our research study is to identify the current and possible technology-driven enhancement of the value chain in the SCFHS.

One of the most significant aspects of the literature is related to the **platform ecosystem**. A converging common line of research contributions covers the need to design and utilize emerging platform ecosystems where the various stakeholders will synch to contribute collaboratively to value delivery. In their recent research titled *"The digital transformation of the healthcare industry: exploring the rise of emerging platform ecosystems and their influence on the role of patients,"* Hermes et al. (2020) highlighted the significance of digital platforms that leverage platform = mediated ecosystems for value cocreation with expanded networks of stakeholders, intermediaries, and providers. In such extended value space, it is important to understand the emerging changing roles, the necessity to design and utilize systems of systems, and the interactions and value delivery across and beyond the boundaries of the participating entities in the ecosystems. In an organization like SCFHS, the understanding of a platform-mediated ecosystem of value cocreation in line with the six core functions is a critical success factor for the implementation of the digital transformation strategy.

In the same direction, Han et al. (2020) elaborated on the concept of disruptive innovation through digital transformation. They concluded that the parties in the current healthcare ecosystem have been impacted to some extent by the complementary functions that emerging managed service providers (MSPs) bring about. They also emphasize the long-term impact of digital transformation due to the evolution of new technologies. Klinker et al. (2020) provided more insights in this context by discussing the role of augmented reality as a digital transformation enabler.

Tables 2 and 3 summarize the key theoretical contributions into our research design that are presented in Section 3.

In this section, we summarized the selected literature to identify significant aspects of the digital transformation strategy. In the next section, we introduce an initial strategic framework for digital transformation in SCFHS that can be used as an international benchmark for healthcare organizations interested in implementing a research-based digital transformation strategy.

TABLE 3 Key interpretation of findings for health workforce planning.

Author	Key emphasis				
	Digital transformation	Health	Ethics	Innovation	Digital technologies
Meskó et al. (2017)	X Digital nudging	X Hospitals			
Müller (2021)	X Telemedicine	X Health services	x		
Meskó et al. (2017)	X Digital health	X Medical technology self-tracking			
Herrmann et al. (2018)	x	X Healthcare sector Healthcare reform		X Incremental innovation Disruptive innovation Organizational innovation	
Benjamin and Potts (2018)	X Digital health	x			
Fernández et al. (2020)	X ICT	x	x		
Jahankhani and Kendzierskyj (2019)	x	X Health information exchange	X Cyber attack		X Blockchain data breaches Patient-centric data
Natsis et al. (2020)	x	X Hospital websites			x e-health
Hasan et al. (2020)	x	X Mental illness			X Digital interventions
Woodside and Amiri (2018)	x	X Healthcare organizations			X Hyperledger blockchain
Hermes et al. (2020)	x	X Healthcare Digital health			X Platform ecosystem Health information technology

Continued

C. Global experiences and approaches to digital transformation in health

TABLE 3   Key interpretation of findings for health workforce planning—cont'd

Author	Key emphasis				
	Digital transformation	Health	Ethics	Innovation	Digital technologies
Han et al. (2020)	x	X Healthcare industry			X e-Health platforms
Peek et al. (2020)	x	X Covid-19			X Digital health technologies
Klinker et al. (2020)	x	X Healthcare			X Augmented reality Smart devices
Warraich et al. (2018)	x	X Healthcare	x		
Frennert (2021)	x	X Elder care			X Welfare technology

3. A strategic framework for digital transformation in Saudi Commission for Health Specialties

One of the key propositions of our research on digital transformation is related to an integrated approach for the design and implementation of a digital transformation strategy in the SCFHS aiming to promote the efficiency and performance of the organization. Our approach combines two significant dimensions of value proposition, namely functional and strategic components (see Fig. 1).

From one side, we are dealing with the **functional aspects** of digital transformation in the SCFHS. There are organized around three significant pillars:

- Business process reengineering (data, knowledge, and process management).
- Technology-driven value proposition for digital transformation.
- Quality assurance (KPIs and benchmarks).

A more detailed overview of these components is provided in Fig. 2.

From the other side, our approach also emphasizes the **strategic components and impact dimensions of the digital transformation** strategy as these are derived from the state of the art of the literature and the recent developments in this scientific domain. These components are summarized as follows:

Strategic components:

- Policy recommendations and strategic implications.
- Digital transformation and digital innovation foundations.
- Value creation and cocreation.

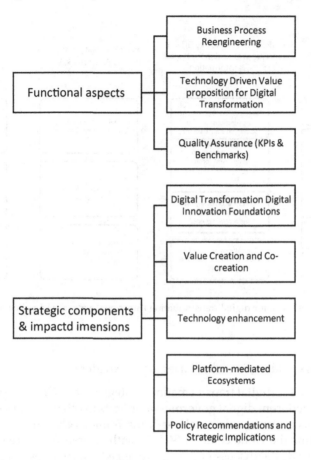

FIG. 1   Functional and strategic components and impact dimensions of digital transformation. *Source: The authors.*

- Technology enhancement.
- Platform-mediated ecosystems.

The quest for **digital transformation and digital innovation** requires a multidimensional approach that integrates complimentary value-adding components. The adoption of digital transformation and digital innovation in the SCFHS needs to work in parallel areas and to provide a unified strategy. Within this context, the following aspects highlighted in the literature review provide significant areas of study in our strategic framework for digital transformation:

- Lessons learned from digital transformation in government.
- Achieve sustainable change at a scale.
- Shift from the traditional healthcare value chain to the Industry 4.0 digital healthcare value chain.
- Transform value proposition, value capture, and value delivery in the healthcare industry.

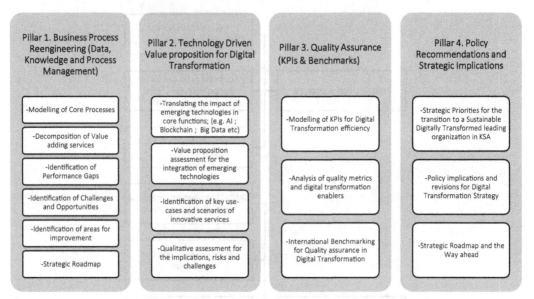

FIG. 2   A strategic framework for digital transformation (functional aspects and policy recommendations).

- Business models.
- Networks of stakeholders, intermediaries, and providers.

The justification of the digital transformation strategy in the SCFHS core functions needs to exploit lessons learned from digital government projects in other areas than healthcare and to interpret the various metaphors in a value-adding framework. Furthermore, the critical requirement to redefine the value chain of SCFHS with reference to core value functions and support functions leads the discussion for an advanced mapping of stakeholders and a strategic analysis of their requirements and cocreation modes for value creation (Fig. 3).

The focus on **value creation and cocreation capability** within the SCFHS is a bold component of our research strategy that will be communicated with more detail in the next section. The determinants of value and the modeling of the value network within the SCFHS and its

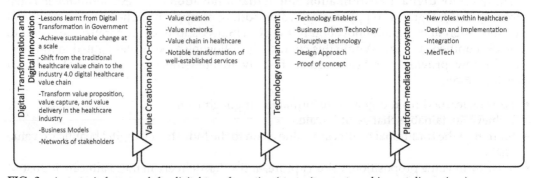

FIG. 3   A strategic framework for digital transformation (strategic aspects and impact dimensions).

extended form that also includes stakeholders and other key institutions in the country are key priorities. This effort will require significant research work in the following areas:

- Value creation
- Value networks
- Value chain in healthcare
- Notable transformation of well-established services

The effective delivery of the previously mentioned research areas will lead to two bold foundations: (i) the justification of notable transformations for enhanced value delivery of well-established services in the core functions of the SCFHS, and (ii) the identification of novel interventions powered by the disruptive application of technologies for radical innovations and new value propositions for the entire SCFHS and its stakeholders.

Thus, the **technology enhancement of the core functions** of SCFHS has a strategic orientation. It is not a case of an occasional deployment of new technological arrivals but a strategic response to the challenges and opportunities faced by the SCFHS in our times.

- Technology enablers
- Business-driven technology
- Disruptive technology
- Design approach
- Proof of concept

In this strategic choice, various research streams need to be identified and studied further. An example is the capacity of technology enablers to promote the key aspects of the vison and mission of SCFHS. The embodiment of business-driven technological capabilities requires well-structured design approaches and the development of proof-of-concept enhanced services that can be integrated on a single platform-mediated ecosystem in SCFHS.

The platform-mediated ecosystems in the SCFHS are another critical research area integrated in our study. The development of an ecosystem through an integrated three-tier approach on data, value-adding services, and roles will set up the basis for the digital transformation strategy. Within this priority, the following areas are important and need systematic research:

- New roles and jobs within health specialties
- Design and implementation approach for the platform-mediated environment
- Integration of the ecosystem to other significant ecosystems of Saudi organizations in healthcare

4. A research plan for digital tranformation in SCFHS

The strategic framework for digital transformation is a high-level approach that can be customized and utilized for the justification of a digital transformation strategy in modern healthcare organizations. It can be used as a consulting instrument and as a managerial tool for the design and implementation of the digital transformation strategy. In this section, we present one more complementary use: The exploitation of the strategic framework for digital transformation for a research project within the SCFHS aiming to provide research-based evidence for the enhancement of the organization.

4.1 A research project for digital transformation in SCFHS

We are implementing a research project in SCFHS aiming to promote digital transformation within the organization. The project's objectives are:

- To examine how business process modeling and reengineering can identify areas of improvement in the core functions of the SCFHS and to justify digital transformation interventions.
- To develop theoretical models and frameworks for technology-driven value proposition in the context of the core functions of the SCFHS and to test them.
- To design quality frameworks including key performance indicators and quality metrics for the implementation and measurement of the execution of the digital transformation strategy.
- To identify the data and process ecosystem in the SCFHS that will enable a platform-mediated integrated approach for digital transformation in the SCFHS.
- To justify policy-making recommendations for a sustainable digitally transformed SCFHS.
- To identify use cases for digital innovation in SCFHS enabled by emerging technologies such as artificial intelligence, blockchain, cloud computing, big data analytics, and metaverse.

In the next subsections, we elaborate on the research questions, for which detailed research tools will be developed with a subsequent collection of empirical data.

The research questions provide a high-level agreement for the depth of the research study execution. They also summarize the main research focus areas and integrate empirical data collection methods. We provide a short paragraph on the key aspect and rationale for each research question and briefly present the key aspects.

4.2 Research question 1. Business process modeling and reengineering for sustainable digital transformation

The value delivery in the SCFHS is organized around well-defined business processes. The detailed modeling and measurement of current efficiency compared to international benchmarks will set the basis for the contextual utilization of technologies that enable digital transformation. Through this perspective, digital transformation in the SCFHS is contextualized and addresses the key challenges associated with the value carriers, namely the business processes.

The following subquestions provide an introductory approach to our proposed research design.

- What are the main services delivered in the six core business functions of the SCFHS?
- What are the business process models of core functions in SCFHS?
- How can we identify the value propositions and value carriers in the services of the SCFHS?

- What are the most significant performance gaps in the core functions of the SCFHS?
- What are the challenges and opportunities faced in the core functions of the SCFHS?
- What are the most important areas for improvement through digital innovations and digital transformation services?
- What are the key aspects and value pillars of a strategic roadmap for the delivery of digital transformation in SCFHS toward sustainability and improved efficiency and performance?

4.3 Research question 2. Technology-driven value proposition for digital transformation of SCFHS core functions

The emerging technologies of our times provide a value mix for strategizing their deployment in the digital transformation strategy of a sophisticated organization such as the SCFHS. The detailed conceptualization of use cases for the design and implementation of value-adding digital services is a key priority in our research. The interpretation of the added value of technologies such as AI, blockchain, cloud computing, big data analytics, and metaverse in the context of the core functions of the SCFHS is a critical requirement for the formulation of the digital transformation strategy of the SCFHS.

The following research subquestions provide further insights for our planned research strategy.

- How can we translate the impact of emerging technologies listed below in the core functions of the SCFHS?
  - Artificial intelligence
  - Blockchain
  - Big data analytics and dashboards
  - Cloud services
  - Metaverse
  - Modeling of core processes in SCFHS
- What is the value proposition assessment for the integration of emerging technologies in the core functions of the SCFHS?
- How can we identify significant, meaningful, and representative use cases and scenarios of innovative services moving forward the digital transformation strategy in the core functions of the SCFHS?
- How can we assess the implications, risks, and challenges of integrating technology-driven innovation in the SCFHS?
- Which are the critical success factors for the introduction of technology-driven innovation in SCFHS?

4.4 Research question 3. Quality frameworks and KPIs for the implementation and measurement of the execution of the digital transformation strategy

Our research strategy is also emphasizing the quality aspects of the digital transformation strategy in the SCFHS. A systematic research approach is described in the following section and includes key considerations for the key performance indicators and the benchmarks of the quality of the digital transformation strategy. Within this context, key priorities in our research include activities related to the specification and modeling of KPIs for digital transformation efficiency in the country. Furthermore, we intend to analyze and provide research-based evidence for a unified model of quality metrics and digital transformation enablers in the SCFHS. For the delivery of this research work, dedicated work will be also assigned on the international benchmarking for quality assurance and metrics for digital transformation. The following subquestions provide an introductory approach to the measurement of impact and quality of the digital transformation strategy in the SCFHS.

- What are the strategic KPIs to be deployed for the measurement of the digital transformation efficiency?
- What is the set of quality metrics and digital transformation enablers that can be utilized in the SCFHS toward improved performance and efficiency?
- What are the international benchmarks for quality assurance in digital transformation that can be adopted and customized for the needs of the SCFHS?
- What is the best approach to involve key stakeholders of the SCFHS to the quality design and assessment of the digital transformation strategy?

4.5 Research question 4. Platform-mediated integrated approach for the digital transformation in the SCFHS

The successful design and implementation of the digital transformation strategy in the SCFHS in our research strategy is directly linked to a platform-mediated ecosystem of value-adding services and value proposition. From this point of view, the interpretation of the strategy requirements in terms of the platform ecosystem is a significant research priority. Some research methodological aspects are summarized as follows:

- What are the new roles within SCFHS required to utilize the value components of the digital transformation strategy?
- What are the key design principles for the platform-mediated ecosystem?
- What are the significant implementation requirements and critical success factors?
- What is the role of health specialists in the platform-based provision of digital transformation value delivery?

4.6 Research question 5. Policy-making recommendation for a sustainable digitally transformed SCFHS

Our research aims to justify policy recommendations and strategic implications for the efficiency and the performance in the SCFHS in the light of an integrated digital transformation strategy. Thus, our research plan intends to clarify strategic priorities for the transition of SCFHS to a sustainable digitally transformed leading organization. This is a critical milestone and the detailed research plan discussed in the next section provides the details of the implementation. Another significant research aim is also related to the analysis of policy implications and revisions for the digital transformation strategy in SCFHS that will be accompanied by a well-justified strategic roadmap.

The following aspects summarize our research strategy in alignment with this research question:

- What are the strategic priorities for the transition of SCFHS to a sustainable digitally transformed leading organization?
- What are the policy implications and the required policy revisions for an effective digital transformation strategy in SCFHS?
- What are the key stages of a strategic roadmap for the efficiency and sustainability of the digital transformation strategy in the SCFHS?

4.7 Research question 6. Use cases for digital innovation in SCFHS enabled by emerging technologies such as artificial intelligence, blockchain, cloud computing, big data analytics, and metaverse

Our research aims to identify significant value-adding interventions in the core functions of the SCFHS to support the integrated digital transformation strategy. From this point of view, one of the most critical challenges in our research is the specification of the use cases that will be identified in the six core functions of the SCFHS. This is a bold response to the challenges faced in the organization and requires a sophisticated methodological approach for collecting information and knowledge from various stakeholders. Within this context, the following questions provide a short list of key priorities in this research direction.

- What are the required data and knowledge for the detailed description of value-adding use cases and scenarios of use for digital transformation services in SCFHS?
- What are the key priorities in each core function of SCFHS for the design of digital services and digital innovations?
- How can the pool of services that will be prioritized be integrated in a fully functional platform-mediated ecosystem of digital transformation in SCFHS?

5. Research challenges, risks, and milestones for the digital transformation strategy in SCFHS

The research priorities discussed in the previous section require a thorough understanding of the associated research challenges and milestones. In Table 4, we elaborate briefly.

TABLE 4   Research challenges, risks, and milestones in digital transformation research.

Research priority	Challenges and risks	Milestones
Business process modeling and reengineering for sustainable digital transformation	• Complicated business models of core functions in SCFHS • Need to collect diverse data with a variety of methods • Understanding of current performance gaps	• Development of a trusted, transparent method for business process modeling • Identification of areas for improvement

Continued

TABLE 4 Research challenges, risks, and milestones in digital transformation research—cont'd

Research priority	Challenges and risks	Milestones
Technology-driven value proposition for digital transformation of SCFHS core functions	• The variety of emerging technologies • The interpretation of the value proposition theory in the context of the SCFHS core functions	• Matching of technologies to areas of improvement • Detailed specification of technology-driven scenarios for digital transformation in SCFHS
Quality frameworks and KPIs for the implementation and measurement of the execution of the digital transformation strategy	• Design of a quality measurement instrument • Difficulty in identifying international benchmarks • Understanding the thresholds for international benchmark measures	• Build a concrete set of specific and high-level benchmarks • Develop a unique scale for measuring quality • Link benchmarks to policy-making guidelines for digital transformation in SCFHS
Platform-mediated integrated approach for digital transformation in the SCFHS	• The need to involve and engage all the stakeholders of SCFHS in the process • Challenge to identify all value-adding services that need to be transformed or enhanced	• Analysis of the impact of platform-mediated ecosystem on the performance of the SCFHS • Design principles for the platform-mediated ecosystem in SCFHS
Policy-making recommendation for a sustainable digitally transformed SCFHS	• Challenge to integrate diverse aspects and strategic views to a unified digital transformation strategy • Difficulty to understand stakeholders' priorities • Need to study the possible resistance to change due to digital transformation strategy	• Update on current policies and procedures in the SCFHS • High-level policies for the evolution of the digital transformation strategy in SCFHS
Use cases for digital innovation in SCFHS enabled by emerging technologies such as Artificial Intelligence, Blockchain, Cloud computing, Big Data analytics, and Metaverse	• Need to collect various data from different stakeholders • Difficulty to prioritize areas of improvements through digital innovation and transformation • Need to identify the synergies between emerging technologies for meaningful innovations in core functions of the SCFHS • Challenge to think about the disruptive character of the emerging technologies for the core functions	• A compact set of well-defined use cases and scenarios for value-adding digital services in core functions of SCFHS • A value matrix for the value proposition of the digital transformation strategy in SCFHS

6. Conclusion

In this chapter, we introduced a strategic framework for the design and implementation of a robust digital transformation strategy in modern healthcare organizations. Insights from the SCFHS were also integrated to highlight the critical aspects of a complicated intellectual effort to mobilize diverse sociotechnical, functional, and strategic components for the realization of the digital transformation strategy. The main contribution of the chapter is summarized in the core parts of the strategic framework that include:

Functional aspects

- Business process reengineering
- Technology-driven value proposition for digital transformation
- Quality assurance (KPIs and benchmarks)

Strategic components & impact dimensions

- Digital transformation digital innovation foundations
- Value creation and cocreation
- Technology enhancement
- Platform-mediated ecosystems
- Policy recommendations and strategic implications

This can be used as a consulting instrument and as a managerial tool for the design and implementation of the digital transformation strategy. In this chapter, we also presented the exploitation of the strategic framework for digital transformation for a research project within the SCFHS aiming to provide research-based evidence for the enhancement of the organization.

References

Agrawal, A., 2020. Bridging digital health divides. In: Science. 369. American Association for the Advancement of Science, pp. 1050–1052.

Bayram, M., Springer, S., Garvey, C.K., Özdemir, V., 2020. COVID-19 digital health innovation policy: a portal to alternative futures in the making. Omics 24 (8), 460–469 "Mary Ann Liebert, Inc., publishers 140 Huguenot Street, 3rd Floor New … ".

Benjamin, K., Potts, H.W.W., 2018. Digital Transformation in Government: Lessons for Digital Health? SAGE Publications Sage UK, London, England.

Bousquet, J., Anto, J.M., Bachert, C., Haahtela, T., Zuberbier, T., Czarlewski, W., Bedbrook, A., Bosnic-Anticevich, S., Walter Canonica, G., Cardona, V., 2021. ARIA digital anamorphosis: digital transformation of health and care in airway diseases from research to practice. In: Allergy. 76. Wiley Online Library, pp. 168–190.

Dugstad, J., Eide, T., Nilsen, E.R., Eide, H., 2019. Towards successful digital transformation through co-creation: a longitudinal study of a four-year implementation of digital monitoring technology in residential care for persons with dementia. In: BMC Health Serv. Res. 19. Springer, pp. 1–17.

Fernández, A., Beratarrechea, A., Rojo, M., Ridao, M., Celi, L., 2020. Starting the path of digital transformation in health innovation in digital health: conference proceeding. Cienc. Innov. Salud 74, 68. NIH Public Access.

Frennert, S., 2021. Hitting a moving target: digital transformation and welfare technology in Swedish municipal eldercare. Disabil. Rehabil. Assist. Technol. 16 (1), 103–111. Taylor & Francis.

Haggerty, E., 2017. Healthcare and digital transformation. In: Netw. Secur. 8. Elsevier, pp. 7–11.

Han, X., Wu, Y., Zheng, J., 2020. Emerging E-Health MSPs in China, Disruptive Innovation through Digital Transformation. Springer, pp. 49–67.

Hasan, M.R., Kayan, S., Kaliyadan, S.U., Hoque, M.R., 2020. Mobile health app for university students with mental illness: a digital transformation design approach from Australia. In Proceedings of the 28th European Conference on Information Systems (ECIS), an AIS Conference. June 15–17.

Hermes, S., Riasanow, T., Clemons, E.K., Böhm, M., Krcmar, H., 2020. The digital transformation of the healthcare industry: exploring the rise of emerging platform ecosystems and their influence on the role of patients. Bus. Res. 13 (3), 1033–1069. Springer.

Herrmann, M., Boehme, P., Mondritzki, T., Ehlers, J.P., Kavadias, S., Truebel, H., 2018. Digital transformation and disruption of the health care sector: internet-based observational study. J. Med. Internet Res. 20 (3), e104. https://doi.org/10.2196/jmir.9498. PMID: 29588274; PMCID: PMC5893888.

Jahankhani, H., Kendzierskyj, S., 2019. Digital transformation of healthcare. In: Blockchain and Clinical Trial. Springer, pp. 31–52.

Klinker, K., Wiesche, M., Krcmar, H., 2020. Digital transformation in health care: augmented reality for hands-free service innovation. Inf. Syst. Front. 22 (6), 1419–1431. Springer.

Lytras, M.D., Visvizi, A., 2018. Who uses smart city services and what to make of it: toward interdisciplinary smart cities research. Sustainability 10, 1998. https://doi.org/10.3390/su10061998.

Lytras, M., Raghavan, V., Damiani, E., 2017. Big data and data analytics research: from metaphors to value space for collective wisdom in human decision making and smart machines. Int. J. Semant. Web Inf. Syst. 13 (1), 1–10 (2017).

Lytras, M.D., Aljohani, N., Daniela, L., Visvizi, A., 2019a. Cognitive Computing in Technology-Enhanced Learning. IGI Global, Hershey, PA, pp. 1–350 2019 doi:10.4018/978-1-5225-9031-6.

Lytras, M.D., Visvizi, A., Sarirete, A., 2019b. Clustering smart city services: perceptions, expectations, responses. Sustainability 11 (6), 1669 2019.

Merolli, M., 2021. Insights from Public Health Researchers into the Digital Transformation of an Educational Lifestyle Course. IOS Press.

Meske, C., Amojo, I., Poncette, A.-S., Balzer, F., 2019. The potential role of digital nudging in the digital transformation of the healthcare industry. In: International Conference on Human-Computer Interaction. Springer, pp. 323–336.

Meskó, B., Drobni, Z., Bényei, É., Gergely, B., Győrffy, Z., 2017. Digital health is a cultural transformation of traditional healthcare. Mhealth 3, 38. https://doi.org/10.21037/mhealth.2017.08.07. PMID: 29184890; PMCID: PMC5682364.

Müller, M., 2021. The Digital Transformation of Healthcare: Exploring Stakeholder Perspectives and Ethical Issues of Digital Technologies in Patient Treatment. Dissertation. https://dspace.ub.uni-siegen.de/bitstream/ubsi/1856/5/Dissertation_Marius_Mueller.pdf.

Natsis, C., Chrysanthopoulos, S., Laladakis, I., Nagkoulis, I., Stamouli, M.-A., 2020. Digital transformation of the health sector in Greece: evaluation of the websites of Health Organizations in the Region of Attica. Open J. Soc. Sci. 8 (11), 65. Scientific Research Publishing.

Peek, N., Sujan, M., Scott, P., 2020. Digital health and care in pandemic times: impact of COVID-19. BMJ Health Care Inform. 27 (1), e100166. BMJ.

Seoane, F., Traver, V., Hazelzet, J., 2021. Value-driven digital transformation in health and medical care. In: Fernandez-Llatas, C. (Ed.), Interactive Process Mining in Healthcare. Health Informatics. Springer, Cham. https://doi.org/10.1007/978-3-030-53993-1_2.

Steinhauser, S., 2021. COVID-19 as a driver for digital transformation in healthcare. In: Digitalization in Healthcare. Nature Publishing Group, p. 93.

Walsh, M.N., Rumsfeld, J.S., 2017. Leading the Digital Transformation of Healthcare: The ACC Innovation Strategy. American College of Cardiology Foundation, Washington, DC.

Warraich, H.J., Califf, R.M., Krumholz, H.M., 2018. The digital transformation of medicine can revitalize the patient-clinician relationship. NPJ Digit. Med. 1 (1), 1–3. Nature Publishing Group.

Woodside, J.M., Amiri, S., 2018. Healthcare hyper chain: digital transformation in the healthcare value chain. Twenty-fourth Americas Conference on Information Systems. New Orleans. https://web.archive.org/web/20220803143507id_/https://aisel.aisnet.org/cgi/viewcontent.cgi?article=1578&context=amcis2018.

Yamamoto, S., 2021. e-Healthcare service design using model-based jobs theory. In: Architecting the Digital Transformation. Springer, pp. 383–395.

Yap, K.Y.-L., Liu, J., Franchi, T., Agha, R.A., 2021. The launch of the International Journal of Digital Health: ensuring digital transformation in healthcare beyond Covid-19. In: Int. J. Digit. Health. 1. IJS Press.

Yip, W., Fu, H., Chen, A.T., Zhai, T., Jian, W., Xu, R., Pan, J., Hu, M., Zhou, Z., Chen, Q., 2019. 10 years of health-care reform in China: progress and gaps in universal health coverage. In: Lancet. 394. Elsevier, pp. 1192–1204.

11

Digital transformation from a health professional practice and training perspective

Abdulrahman A. Housawi[a] and Miltiadis D. Lytras[b]

[a]Saudi Commission for Health Specialties, Ministry of Health, Riyadh, Saudi Arabia
[b]College of Engineering, Effat University, Jeddah, Saudi Arabia

1. Digital transformation (DT)

1.1 Digital transformation and the fourth industrial revolution

Since the advent of the Fourth Industrial Revolution (Kagermann, 2015) earlier this century, technology has permeated all aspects of life and impacted society at various levels: cultural, geographical, and economic. Industry 4.0 is described by two major paradigms-decentralization and automation (Kagermann, 2015). Decentralization enables the distribution of service provision along different sites and service providers to ensure high flexibility and productivity (Meister et al., 2019, p. 330). This change is associated with an ever-growing complexity. Healthcare systems are not spared, and healthcare costs are staggeringly rising, as is the demand for healthcare and maintenance.

The tremendous advancement in digital technologies is coupled with a remarkable increase in society's dependence on them. This reliance on technology reached a new high with the dramatic arrival of the COVID-19 pandemic two years ago. Countries, communities, and organizations found themselves in a position where there was no option but to adopt technology at rates much faster than before. Ignoring the impact of technology on humanity does not seem to be an option. This would be akin to a journey back to the dark ages. This reality puts a great responsibility on the leadership of organizations to have a clear strategy for how technology will be used to remain relevant. For different complex systems, including healthcare, such an approach is likely to maximize the benefit of adopting technology and bring novel solutions to problems that never existed before.

During the past decade, several definitions for digital transformation have been published. Most describe the transformative effect on businesses driven by recent technological advancements. In many fields, this impact has been revolutionary (Rosalia et al., 2021, p. 1; Iljashenko et al., 2019; Persson et al., 2019).

Rosalia et al. argue that digital transformation could facilitate value-based healthcare (VBHC) through cost-cutting as well as the improved efficiency and productivity of healthcare systems (Rosalia et al., 2021).

The past decade has witnessed a remarkable expansion in digital technologies in the delivery of healthcare and the training of health professionals.

1.2 The intersection of clinical practice/services and health professionals' education/training

One of the realities of the healthcare system is that the training and development of health professionals are intertwined with their clinical practice. As a result, it is nearly impossible to think of health professionals' training outside the context of their clinical practice. As logical as this association is, it brings a dramatic challenge when planning clinical services and training healthcare professionals. This challenge is particularly relevant in situations where the people responsible for each are different entities. Health systems and training leaders are at a tight junction in optimizing training and its results while ensuring the continuity and integrity of clinical practice.

Ironically, there is constant competition between these two domains in most organizations. In such a sibling rivalry, the balance is often tipped in favor of the delivery of clinical services. Solving this problem remains one of the toughest mysteries in healthcare systems worldwide. This is another example where effective leadership in the healthcare system is of critical value in maintaining this demand balance (clinical service-health training/education), which is critical to sustainable success.

2. Challenges and barriers to adopting digital transformation

Compared to other industries, the healthcare industry is relatively slower in its response to technological advancements. There is also significant variability in the extent and speed of digital transformation within different domains of healthcare systems. This variability is attributable to several factors, including: (1) resistance to change from a lack of willingness of practitioners and leaders in the healthcare system to change their practices and adopt new technologies in their work environment, (2) limited familiarity with technology among some healthcare practitioners creates a challenge in technology adoption, (3) the high upfront cost for adopting new technologies and the reluctance of funding agencies to support initiatives with high technology infrastructure requirements (regardless of their potential long-term value), (4) interoperability of new technologies with existing ones, and (5) information privacy and cybersecurity risks (Cenaj, 2021; Velez-Lapão, 2019, p. 1).

2.1 Challenges for transforming the training and healthcare environment

In addition to challenges to adoption, the diversity of stakeholders, specialties, and practices involved within these specialties increases the complexity of digital transformation (DT). This complexity is then reflected in the training of healthcare professionals. It remains unclear how the interaction between health and digital educational transformation will play out at its intersection. However, there is a potential benefit that digital health transformation could play a major role in advancing health professionals' training by improving training capacities. This would allow the opportunity to accommodate more trainees with better training outcomes. We will briefly address these opportunities later in this chapter.

2.2 Opportunities for transforming training in the healthcare environment

If implemented properly, DT in health systems has the additional potential benefit of reforming the training of health professionals. Decentralizing health service provision is likely to create new training venues, and more trainees will have greater (and potentially personalized) clinical training. Decentralized training can also reduce the staggering cost of training health professionals. Other benefits include improved efficiency in training delivery and its quality.

2.3 Examples for integrating technology, training methodology, and clinical practices to advance training objectives

In healthcare, a range of specialties has integrated technology as a key component in training activities at rates dependent on the specialty and the institution, among other factors. Critical care, anesthesia, and emergency medicine are the most common specialties adopting virtual and augmented reality (VR/AR) technologies.

More recently, specialties such as ophthalmology, medical imaging, and pathology have embraced several emerging technologies such as artificial intelligence (AI) and big data analytics. So far, this adoption of technology has led to a transformative advancement in health delivery in those disciplines. These advancements include improved diagnostic accuracy, predictive ability, and access to healthcare.

Experts anticipate faster technology adoption rates during this decade, resulting in a revolutionary advancement in healthcare systems. This advancement warrants an equivalent transformation in health workforce development to optimize the potential social benefits expected from the future health system.

Healthcare system and education practitioners, leaders, and policymakers are responsible for anticipating and leading the DT to maximize its benefit to society.

2.4 Use cases for digital transformation in health training

In this section, we summarize current initiatives in the Saudi Commission for Health Specialties (SCFHS) and its Research and Innovation Center toward DT.

2.4.1 A functional view of the SCFHS

The scope of functional processes engaged by the SCFHS can be summarized by the following listing.

2.4.2 *Matching system for selecting for postgraduate trainees*

The SCFHS has to maintain its integrity and reputation as a regulator for the selection process of postgraduate trainees to begin their journey to become healthcare practitioners. The SCFHS matching system (SCFHS-MS) is a national project that provides an equitable and objective matching system for postgraduate training programs at the SCFHS. The matching system is considered the primary system for applying for Saudi residency and a postgraduate diploma.

The system uses a computerized mathematical algorithm to match or place applicants into the most preferred residency or postgraduate training programs throughout accredited training centers. It requires data points such as the number of available seats based on the accreditation data, the applicant's preferences, the aggregate score, and rank order lists (ROLs) of programs from both the applicant and the training centers.

SCFHS, in coordination with the accreditation data, identifies the number of training seats available for accredited training centers and which programs can participate in the matching process. After applicants submit their personal information and preferences, a special committee validates and aggregates their score based on the following three weighted criteria:

(1) GPA, 30%.
(2) Saudi license examination results, 50%.
(3) Portfolio, 20%.

2.4.3 *Training design, development, and implementation*

(1) **Training** for Postgraduate and Diploma—The primary and major initiatives for training under the SCFHS responsibility that focuses on providing Postgraduate Medical Education (POSTGRADUATE) and Diploma Programs for allied health specialties training in the country. In addition, it provides a mentoring program for trainers through the train-the-trainer initiative (TOT).

    The initiatives include designing/developing, approving, and supervising professional health-related programs as well as developing sustained medical education programs for health-related disciplines in line with the general educational framework. This accounts for more than 80% of the SCFHS's training activities.
- Postgraduate, doctors and specialists (fellowship).
- Diploma, allied health, technicians

(2) The **Health Leadership Academy** was established in 2017 with a focus on training and developing leaders in healthcare institutions at all levels. It also helps them acquire high-level learning experiences to support the National Transformation Program (NTP) in the healthcare system.

(3) **The Health Academy** was also established in 2017 to raise the standards and quality of several healthcare vocational training programs to serve the localization of the health workforce and to increase their efficiency and effectiveness. It includes some allied health areas as vocational learning for the health workforce job market.

(4) **Continuous Professional Development (CPD)**: CPD includes the developmental activities that HCPs undertake through accredited providers to maintain their license to practice with the aim to develop and enhance their knowledge, skills, and attitudes in order to improve the safety and quality of care provided to patients.

(5) **Research and development** (R&D) provides a research mentorship training program for interns/trainees to aid in developing their skills and ability to conduct research.

(6) HR L&D Internal employee learning and development to enhance organization capability and performance.

2.4.4 Admission based on predefined criteria to programs

(1) Postgraduate for university graduates to enter residency programs.
- **(i)** Saudi professional licensure exam for final year health college students
- **(ii)** Matching system
  1. Bachelor's degree GPA, 30%.
  2. Saudi professional licensure exam, 50%.
  3. Portfolio, 20%.
  4. Choice of residency institution.

(2) Allied Health and Diploma—Application?? for Training affairs Saudi Postgraduate Diploma" program" which qualifies its holder to be a Registrar or Senior Specialist in a specialty after completing the required experience.

(3) HLA—As part of the MOH drive to equip current and emerging leaders with the managerial and leadership skills required to support the health transformation.
- **(i)** Application and fees

(4) HA training is based on prerequisites to repurpose nonemployed nationals to participate in the local health sector by increasing the level of knowledge and effectiveness to raise the standards and quality of several healthcare vocational programs. The intent is to contribute to the Health Transformation Program.
- **(i)** Application and fees
- **(ii)** Bachelor's degree
- **(iii)** Not working

(5) CPD—Training programs in line with the practice of HCP skills and ability development.
- **(i)** Approved provider from SCFHS

(6) R&D—Research training admission requirements.
- **(i)** Trainee registration

(7) Employee Learning and Development: To enhance organization capability, capacity and performance based on policy of two trainings per year. Employee needs assessment and suggested development courses.
- **(i)** Employee request and approval

2.4.5 Assessment: Exams and evaluation

The goals of SCFHS assessment initiatives are to assess healthcare practitioners' ability to apply their knowledge, concepts, and principles and determine the fundamentals of patient-centered skills that are important in health and disease management for safe and effective patient care. In addition, it serves as a lever for ensuring that practicing health professionals meet specific standards and continue to maintain competence in a given content area.

One of the central SCFHS goals is to ensure a healthy community through competent healthcare providers based on best standards by protecting and promoting health led by compassionate and competent practitioners.

2.4.6 Classification and registration

This is a scheme of classifying HCPs according to their qualification (education, skills, and experience) and registering them in a unified database as licensed healthcare practitioners in Saudi Arabia.

The process considers Saudi board graduates for allied health, diploma, and postgraduates and preemployment for foreign doctors, specialist, or technicians. The high-level process steps are:

(i) Submit application, fees, documents/certificates, and justification for classification. See regulation for details.
(ii) The following steps are to assess the application as per the classification requirements and regulations and process the outcome as follows:
  • Grant classification based on regulations.
  • Request additional information.
  • Reject the application for not meeting the requirements and regulations.

After classification, the following assessment process is related to registration to validate employment eligibility. The process validates HCP employment eligibility and places them in a unified database to be licensed healthcare practitioners.

Once all the regulatory criteria are met, the registration process assesses the following and issues the license to practice.

(a) Classification is complete and in good standing.
(b) Valid employment letter indicating the clinical practice area of specialization.
(c) Valid identification.
(d) Medical report for applicants over 65 years.

2.4.7 Accreditations

This is to ensure effective systems and infrastructure are in place to deliver the intent of the training. The examination and assessment processes are developed and deployed to determine the competency level of healthcare applicants, trainees, and foreign employees seeking to work in the country. The training department curricula are the essential inputs/drivers to the assessment/exam design processes, whether new training areas/specialties, classification, reclassification, or any new areas of certification from the Health Academy portfolio.

Best practices

World-class entities in Medical licensure have been collaborated with the Commission for the establishment of best processes and automation. These institutions were the National Board of Medical Examiners (NBME) and the National Board of Osteopathic Medical Examiners (NBOME), both from the United States, and the Medical Council of Canada (MCC).

These engagements provided expertise and standardized frameworks for creating customized assessment solutions to meet the SCFHS's medical testing and assessment needs. Areas of focus included item writing and security for written examinations, case designs for clinical skills assessments, and oral examinations. Additionally, physical facilities design, the framework for establishing examination committees, and item-generating software were addressed.

2.4.8 Health workforce planning

This is to make available the right kind of personnel in the right numbers, with appropriate skills at the right place to do the right job.

2.4.9 Continuing professional development

This includes programs to help HCPs develop and update their professional knowledge and skills.

Councils

In the next section, we provide an indicative list of DT projects within the context of the core functions of the SCFHS.

2.4.10 Indicative use cases for functional digital transformation projects in SCFHS

The selection of DT projects within a well-established organization that serves a whole country such as Saudi Arabia needs strategic justification as well as a well-defined execution plan with reference to outcomes and impacts on sustainability and development.

In Table 1, we provide an indicative list of possible projects in the SCFHS. As explained in Chapter 10, a strategic framework can guide its selection and implementation. Each use case

TABLE 1 Indicative digital transformation use cases in the SCFHS.

Functional area	Use cases for digital transformation
Matching system for selecting postgraduate trainees	• Use of machine learning (ML) and artificial intelligence (AI) for utilization of the matching process • Development of an online dashboard for understanding impact of variables on matching process
Training design, development, and implementation	• Use of AR/VR/metaverse for a new medical training experience (Alhalabi and Lytras, 2019) • Use of AI for personalized medical training • Dynamic skills and competencies management for medical know-how transfer • Design of e-learning personalized spaces • Transformation of academia/SCFHS/med tech industry value cocreation
Admission based on predefined criteria to programs	• Use of AI and ML for automatic admission • Recommendation systems for admission-based decisions • Talent-based admission management
Assessment: Exams and evaluation	• Team-based assessments • Use of AI for enhancement of the quality and objectivity of the assessment
Classification and registration	• Use of ML for automatic classification
Accreditations	• Reengineering of accreditation procedures • Use of key performance indicators (KPIs) and big data analytics

Continued

TABLE 1   Indicative digital transformation use cases in the SCFHS—cont'd

Functional area	Use cases for digital transformation
Health workforce planning	• Innovative training programs for health specialties of the future • Dynamic agoras of skills and competencies
Continuing professional development	• Massive open online courses for CPD • AR/VR/metaverse for CPD
Councils	• Value cocreation

requires systematic study and deep analysis of technical, functional, and strategic requirements. It is beyond the scope of this chapter to communicate these parameters.

2.4.11 Digital transformation project case: Value cocreation with stakeholders in SCFHS

One more example of a DT project in the SCFHS is related to the utilization of its stakeholders toward enhanced performance and impact. The SCFHS is involved in strategic partnerships with its stakeholders. The current agreements cover a variety of initiatives and research development and research partnership agreements. Within the scope of our approach, we intend to assess the current utilization and impact of these agreements as well as understand how the value cocreation process enhances and promotes the strategic character of the partnership and how this process can be digitally transformed.

Overview of current partnerships and agreements

Overall, four different types of partnerships and stakeholder relations are identified:

- Health transformation support.
- Enabling health transformation.
- Research development and research partnerships.
- Supporting research creativity.

Each of these types of collaborations plays a significant role in the promotion of the strategic objectives of the SCFHS. In our research, we are interested in understanding these types of partnerships in terms of added value, utilization and impact to sustainability, innovation, and partnership. Beyond this significant objective, we also need to investigate new forms of partnerships and value cocreation that will involve more stakeholders currently not in the portfolio of partners for SCFHS. This will lead to a new era of service and impact of the SCFHS in Saudi Arabia. Table 2 provides an initial categorization of different types of stakeholders that are collaborating currently with SCFHS. The current portfolio of agreements is related to 14 different types of stakeholders:

- Affiliated universities and research centers.
- Educational and training centers.
- Health associations.
- Health councils and committees.

TABLE 2   Stakeholders and types of agreements.

Type of stakeholder	Number of agreements
Affiliated universities and research centers	13
Hospitals, communities, and health centers	13
Educational and training centers	8
Ministries	7
Research centers and institutions	6
Health associations	5
Health councils and committees	5
Organizations	5
Royal colleges and the like	5
Supportive research services	5
NCs	3
Libraries	2
Other	2
Scientific publishing houses	1

- Libraries.
- Ministries.
- Organizations.
- Research centers and institutions.
- Royal colleges and the like.
- Scientific publishing houses.
- Supportive research services.
- Hospitals, communities, and health centers.
- NCs.
- Other.

In our research, we intend:

- To understand the level of current utilization of agreement for each different type of stakeholder.
- To revise the strategic focus and intended impact of the agreement with different types of stakeholders.
- To develop a masterplan for strategic utilization of stakeholder portfolio.
- To consult with different stakeholder entities to define and to agree to a new value cocreation and value realization strategy.
- To define metrics and quality KPIs for monitoring and measurement of the efficiency of the collaboration and partnership.
- To prioritize the strategic objectives under each type of stakeholder.

Table 2 provides an overview of current agreements per different type of stakeholder.

The research promotes an integrated approach for the design and implementation of an integrated strategy in the SCFHS aiming to promote the utilization of stakeholders and their involvement in joint initiatives with significant impact in the performance of the organization and the realization of bold value proposition.

Our approach combines two significant dimensions of stakeholder theory. From one side, we are dealing with the **identification and utilization of stakeholders** in the SCFHS. This effort is organized around two significant pillars:

Functional components
− Identification of primary and secondary stakeholders in SCFHS.
− Assessment of current utilization of partnership of SCFHS with existing stakeholders.

From the other side, our research proposal also emphasizes the **strategic impact of the stakeholder theory and its capacity to enhance the developmental research-driven performance in the SCFHS.**
These components are summarized as follows:

Strategic components
− Reinforcement strategy for developmental research with stakeholders.
− Quality management and continuous improvement.
− Developmental research projects roadmap for stakeholders.
− Digital hub and digital transformation of stakeholder network.

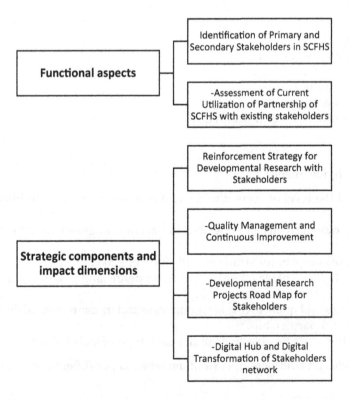

C. Global experiences and approaches to digital transformation in health

The study's objectives are:

Research core part:
- To deliver a systematic research approach for the identification of primary and secondary stakeholders in SCFHS with an emphasis on common interests, challenges, and opportunities.
- To develop methodological instruments and research tools for the assessment of current utilization of partnership of SCFHS with existing stakeholders, both primary and secondary.
- To provide research-based evidence for a reinforcement strategy for developmental research with stakeholders in SCFHS.
- To justify value cocreation Strategies for the SCFHS and its stakeholders based on common interests, priorities, and strategic agreements for joint value proposition.
- To apply total quality management and continuous improvement in the strategic approach for stakeholder utilization toward increased efficiency and performance in the SCFHS.

Developmental core part:
- To identify and to prioritize the components of a Developmental research projects roadmap for stakeholders value cocreation in the SCFHS.
- To set up the phase 1 of the digital hub of stakeholders value cocreation network.
- To justify policy-making recommendations for the utilization of the impact and the SCFHS.

2.5 Conclusions

The design and implementation of DT projects in the context of health education and training need to consider various factors, including business, human, social, and technological factors (Chui and Lytras, 2019; Daud et al., 2019; Guzmán et al., 2018; Lytras and Chui, 2019; Lytras and Visvizi, 2019; Lytras et al., 2018, 2019a, 2019b, 2019c; Zhang et al., 2018; Zhang et al., 2019). It also has to interpret the value proposition of emerging technologies in the context of the current functional and strategic challenges of health training institutions. Within this context in this chapter, we tried to provide a systematic approach on the interpretation of challenges and barriers to feasible, sustainable DT projects. The implementation of DT projects requires a deep understanding of the strategic implications and the impact in terms of value creation and performance. This chapter can guide policy makers, computer scientists, educators, physicians, and consultants to develop a common language and a common understanding for the challenges of DT within modern healthcare organizations.

References

Alhalabi, W., Lytras, M.D., 2019. Editorial for special issue on virtual reality and augmented reality. Virtual Real. https://doi.org/10.1007/s10055-019-00398-6.

Cenaj, T., 2021. COVID-19 and the digital transformation of health care. Telehealth Med. Today 6 (1). https://doi.org/10.30953/tmt.v6.243.

Chui, K.T., Lytras, M.D., 2019. A novel MOGA-SVM multinomial classification for organ inflammation detection. Appl. Sci. (Switzerland) 9 (11), 2284. https://doi.org/10.3390/app9112284.

Daud, A., Lytras, M.D., Aljohani, N.R., Abbas, F., Abbasi, R.A., Alowibdi, J.S., 2019. Predicting student performance using advanced learning analytics. 26th International World Wide Web Conference 2017, WWW 2017 Companion. pp. 415–421. doi:10.1145/3041021.3054164.

Guzmán, G., Torres-Ruiz, M., Tambonero, V., Lytras, M.D., López-Ramírez, B., Quintero, R., Moreno-Ibarra, M., Alhalabi, W., 2018. A collaborative framework for sensing abnormal heart rate based on a recommender system: semantic recommender system for healthcare. J. Med. Biol. Eng. 38 (6), 1026–1045. https://doi.org/10.1007/s40846-018-0421-y.

Iljashenko, O., Bagaeva, I., Levina, A., 2019. Strategy for establishing personnel KPI in digital transformation of health care organizations. IOP Conf. Ser.: Mater. Sci. Eng. 497 (1), 12029. https://doi.org/10.1088/1757-899x/497/1/012029.

Kagermann, H., 2015. Change through digitization—value creation in the age of industry 4.0. In: Albach, H., Meffert, H., Pinkwart, A., Reichwald, R. (Eds.), Management of Permanent Change. Springer Gabler, Wiesbaden. https://doi.org/10.1007/978-3-658-05014-6_2.

Lytras, M.D., Chui, K.T., 2019. The recent development of artificial intelligence for smart and sustainable energy systems and applications. Energies 12 (16). https://doi.org/10.3390/en12163108. art. no. 3108.

Lytras, M.D., Visvizi, A., 2019. Big data and their social impact: preliminary study. Sustainability (Switzerland) 11 (18), 5067. https://doi.org/10.3390/su11185067.

Lytras, M.D., Aljohani, N.R., Visvizi, A., Ordonez De Pablos, P., Gasevic, D., 2018. Advanced decision-making in higher education: learning analytics research and key performance indicators. Behav. Inform. Technol. 37 (10–11), 937–940. https://doi.org/10.1080/0144929X.2018.1512940.

Lytras, M., Visvizi, A., Damiani, E., Mathkour, H., 2019a. The cognitive computing turn in education: prospects and application. Comput. Hum. Behav. 92, 446–449. https://doi.org/10.1016/j.chb.2018.11.011.

Lytras, M.D., Visvizi, A., Sarirete, A., 2019b. Clustering smart city services: perceptions, expectations, responses. Sustainability (Switzerland) 11 (6) art. no. 1669 doi:10.3390/su11061669.

Lytras, M.D., Chui, K.T., Visvizi, A., 2019c. Data analytics in smart healthcare: the recent developments and beyond. Appl. Sci. (Switzerland) 9 (14) art. no. 2812 doi:10.3390/app9142812.

Meister, S., Burmann, A., Deiters, W., 2019. Digital health innovation engineering: enabling digital transformation in healthcare: introduction of an overall tracking and tracing at the Super Hospital Aarhus Denmark. In: Digitalization Cases, pp. 329–341.

Persson, M., Giunti, G., Grundstrom, C., 2019. Customer attitudes towards participation and health data sharing in the digital transformation of Finnish insurance. In: Bled eConference, p. 18, https://doi.org/10.18690/978-961-286-280-0.43.

Rosalia, R.A., Wahba, K., Milevska-Kostova, N., 2021. How digital transformation can help achieve value-based healthcare: Balkans as a case in point. Lancet Regional Health-Europe 4, 100100. https://doi.org/10.1016/j.lanepe.2021.100100.

Velez-Lapão, L., 2019. The challenge of digital transformation in public health in Europe. Eur. J. Pub. Health 29. https://doi.org/10.1093/eurpub/ckz185.180.

Zhang, X., Zhang, Y., Sun, Y., Lytras, M., Ordonez de Pablos, P., He, W., 2018. Exploring the effect of transformational leadership on individual creativity in e-learning: a perspective of social exchange theory. Stud. High. Educ. 43 (11), 1964–1978. https://doi.org/10.1080/03075079.2017.1296824.

Zhang, X., Lytras, M.D., Aljohani, N.R., 2019. Cognitive computing alternate research track chairs' welcome. 26th International World Wide Web Conference 2017, WWW 2017 Companion. pp. 1–3.

Further reading

Arafat, S., Aljohani, N., Abbasi, R., Hussain, A., Lytras, M., 2019. Connections between e-learning, web science, cognitive computation and social sensing, and their relevance to learning analytics: a preliminary study. Comput. Hum. Behav. 92, 478–486. https://doi.org/10.1016/j.chb.2018.02.026.

Chui, K.T., Liu, R.W., Lytras, M.D., Zhao, M., 2019. Big data and IoT solution for patient behaviour monitoring. Behav. Inform. Technol. 38 (9), 940–949. https://doi.org/10.1080/0144929X.2019.1584245.

12

Research and Education Skills as a core part of Digital Transformation in Healthcare in Saudi Arabia

Basim Saleh Alsaywid[a,b], Jumanah Qedair[c], Yara Alkhalifah[d], and Miltiadis D. Lytras[e,f]

[a]Department of Urology, Pediatric Urology Section, King Faisal Specialist Hospital and Research Centre, Riyadh, Saudi Arabia [b]Education and Research Skills Directory, Saudi National Institute of Health, Riyadh, Saudi Arabia [c]College of Medicine, King Saud bin Abdulaziz University for Health Sciences, Jeddah, Saudi Arabia [d]Fresno College of Public Health, California State University, Long Beach, CA, United States [e]College of Engineering, Effat University, Jeddah, Saudi Arabia [f]Distinguished Scientists Program, Faculty of Computing and Information Technology, King Abdulaziz University, Jeddah, Saudi Arabia

1. Introduction

Digital transformation in Saudi Arabia began in 2018 with the implementation of Vision 2030's rapid digital change. The systems focus on the Saudi eHealth Exchange (SeHE), telemedicine, cybersecurity, system certification, advanced functionality, artificial intelligence, analytics, blockchain, and models of care. In 2020, the digital transformation plan updated the mission as follows: "To transform healthcare delivery through technology to deliver safer, more efficient healthcare services for the population of Saudi Arabia." It also encouraged the use of a digital health national transformation program (2018).

In 2020, a major digital response emerged during the COVID-19 pandemic in Saudi Arabia. Telemedicine has enabled some telehealth apps such as Sehaty, Tawaklna, and Tataman. The Ministry of Health (MoH) changed its mission to prevention with a focus on preventing noncommunicable diseases. With all these enormous digital transformation changes in the Saudi Arabia health sector, there is still a need to enable more digital health solutions.

The aim of this chapter is to state the requirements needed for the transition to e-health in times of artificial intelligence by stating the major challenges and discussing the issues faced by the Saudi health system including data, electronic records, and clinical issues. Then we propose an action plan including initiatives and suggestions for future digital health transformation through digital health solutions and discuss the future in healthcare with new technology.

All these transitions and issues for digital transformation are then discussed in detail as an organizational developmental process by utilizing the strategy of transformation. Finally, the authors discuss the action plan and milestones for digital transformation in the Saudi National Institute of Health (SNIH) by transforming SNIH into an innovation hub that attracts business innovators.

1.1 Understanding the requirements for the transition to e-health in times of artificial intelligence

In spite of the high potential of deploying artificial intelligence (AI) in the healthcare system, many challenges can impede AI solutions. A group of United Kingdom healthcare professionals and academics discussed the challenges they faced in building AI-based healthcare solutions. Their discussion was organized into four phases: conceptualization, data management, AI application, and clinical application (Wilson et al., 2021).

An essential tool in the healthcare system is electronic health records (EHRs) that electronically collect and store patient data and medical history, facilitating the accessibility and use of these data. Blockchain technology and natural language processing techniques such as machine learning can be utilized to extract a specific type of data from EHRs. This in turn can ease decision making and analysis of medical data. According to Wilson et al. (2021), these are cornerstones of e-health.

In Saudi Arabia, the MoH established a strategy for digitizing the health sector. Many requirements and needs were addressed to ensure the efficiency and effectiveness of the strategy. Setting policies and creating directory committees to monitor the progress is the first step toward the transition to e-health. Also, health data has to be managed by creating an architecture, standardizing its utilization, and allowing interoperability across the MoH enterprise. To optimize the outcomes of the strategy, the MoH developed funding models and investment plans. The value of innovation and digital reinvention was also considered due to its crucial impact on incorporating new digital models and testing new solutions. This opens the doors for research and analysis for further evaluation and advancement of the local strategy according to the digital health strategy.

While many organizations are involved in the e-health strategy, the Saudi government plays the key role in supervising the roadmap of digital health to ensure its alignment with the country's goals and enable accountability across all stakeholders. A highly competent workforce is the engine of this desired transformation. Therefore, empowering youth and recruiting specialists in the areas of project management, cybersecurity, data governance, service design, and informatics are priorities.

To improve data quality in Saudi Arabia, the Saudi Health Informatics Competency Framework (SHICF) was developed based on evidence by expert researchers. It includes six main domains and 22 subdomains demonstrating the important theoretical and practical aspects of health informatics (HI) locally. This framework was proposed by Almalki et al. (2021) to fill the gap in the HI landscape by meeting workforce and market needs. It defines the field of HI in the region and provides an effective tool for evaluating HI candidates and planning HI educational programs (Fig. 1).

1.2 Key milestones of future action plans for digital transformation

In 2020, the MoH updated the mission "to transform healthcare delivery through technology to deliver safer, more efficient healthcare services for the population of Saudi Arabia" with three goals (National Transformation Program., 2018):

- To improve health by increasing the duration, happiness, and quality of life of Saudi residents, with the Vision 2030 target of raising citizen life expectancy to 80 years by 2030.
- To improve healthcare by enhancing the quality and consistency of services as well as the performance and accountability of healthcare organizations and personnel to provide safe, effective, patient-centered, timely, and equitable treatment.
- To increase value through limiting costs, improving outcomes, managing public healthcare spending, and directing new investment.

All three transformation goals are aligned with and support the Vision 2030 strategic health objectives of access, value, and public health.

The transformation aspects are within seven themes:

- New models of care.
- Provider reforms.

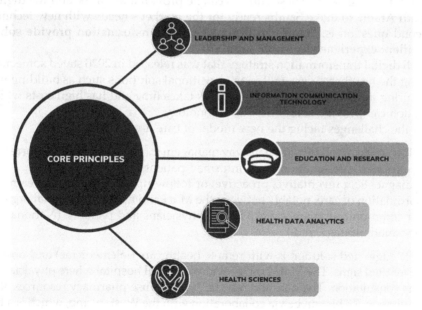

FIG. 1   The Saudi Health Informatics Competency Framework (SHICF) (Almalki et al., 2021).

- Financing reforms.
- Governance development.
- Private and third sector participation.
- Workforce development and e-health development.

Within these themes, the solutions are under three classes: national solutions, cluster solutions, and enterprise solutions. The first question is, "Why do we need a digital strategy transformation?" The need for a digital transformation strategy is to have a healthcare system that is more patient-centered as well as to leverage patient experience with new technology solutions. This will allow the use of innovation technology that assists a large population with less human error.

The question now is, **"What is the first step for the Saudi health digital transformation?"** The answer relies on one of the biggest issues facing most health entities, which is data. The country has already established dedicated institutions, such as the Saudi Data and AI Authority (SDAIA), the National Center for Artificial Intelligence (NCAI), and the National Data Management Office (NDMO), as well as adopted a regulatory framework for the use of robotics and automation (National Data Governance Interim Regulations, Internet of Things (IoT) Regulatory Framework) and SeHe (the Saudi Health information exchange).

The implications for digital health solutions require good quality data; therefore, the first step is to have data management to drive the nation into a data-driven system and to have better data quality for digital health solutions. Huge amounts of data are generated by new technology, which can be evaluated to give real-time clinical or medical care. The fundamental challenge in extracting value from large data is making it relevant, actionable, available, and interoperable. With data management, there is a need for professionals who understand data. Therefore, another step that should be taken is enabling data scientists, data analysts, and data engineers. These courses can be provided by MISK and the digital Academy in Saudi Arabia to make Saudis ready for the market's needs with new technology.

The second question is, **"How can digital health transformation provide solutions to leverage patient experience in Saudi Arabia?"** The Saudi digital transformation strategy that was released in 2020 stated some major challenges facing the healthcare system. Using transitional solutions such as building more hospitals and hiring more doctors in remote regions takes time and has high costs while digital transformation can provide solutions with technology.

Some of the challenges facing the new model of care are:

1. Patient flow is disrupted by poor pathway management, which includes incorrect referrals and inappropriate presentation by ill-informed patients.
2. Lack of diagnostic, preventative, proactive, or follow-up out-of-hospital treatments.
3. Poor coordination of care, notably between the MoH and nongovernmental organizations; and poor communication among providers, physicians and patients. (National Transformation Strategy, n.d.)

A possible suggested solution is with remote health care (telemedicine) and onsite patient monitoring for vital signs. The MoH now provides a virtual hospital where physicians provide telemedicine consultation. The question now is, "How to use pharmacy resources along with virtual consultation"? This can be via collaboration with the Wasfaty app, which is a telehealth

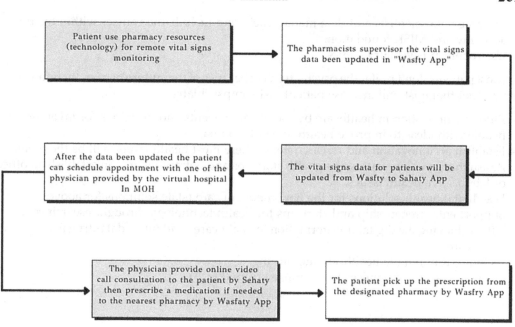

FIG. 2   An example of the interconnection among telehealth applications in the Saudi healthcare system.

application for pharmacy prescriptions by MoH, and the Sehaty application. Sehaty is an MoH application that regulates patient medical records along with appointments and virtual consultation. The patient can use the pharmacy as a place to have vital signs taken and recorded in Wasfaty by the pharmacist. These data can then be transferred into Sehaty where the physician can use them for diagnosis. Then the medication can be prescribed via Wasfaty (see Fig. 2).

Finally, **"What is the next plan after the digital transformation plan?"** The future will be the shift from digital health to data-driven health. The future plan will shift from digital transformation to data-driven transformation. Big data, in particular, is the catalyst for progress in the healthcare industry. Big data analysis combines information and allows for the discovery of patterns and trends. Big data can aid the healthcare business in several ways, including more precise staffing and aiding hospitals in predicting future admission rates.

The facilitation of chronic care creates lean systems for ongoing and standardized treatments, allowing for the successful management of a population risk cohort.

A lower rate of medication errors can detect and indicate any discrepancies between a patient's health circumstances and medicine prescriptions, alerting both health professionals and patients, according to Nico et al. (2022).

Suggested initiatives for a future digital transformation strategy include but are not limited to the following:

– Enable technology for the digital transformation of the Internet of Things and artificial intelligence solutions in Haj to leverage their experience in Haj as well as improve their health during the Haj (Telemedicine and onsite patient monitoring stations, mobile clinic with virtual consultation).

- Promote data scientists and data management and include the courses within the digital academy and MISK foundation.
- Use robots in psychiatry clinics.

Most patients don't prefer the psychiatry visit because of the culture barrier; therefore, having a robot therapist will increase patient visits in psychiatry.

- Promote innovation in healthcare by establishing events and programs for inventors and people with ideas to improve health in Saudi Arabia.
- Establish an innovation and research park where most health issues will be the theme. Researchers and innovators along with start-ups and entrepreneurs can establish an office or lab to solve health issues.
- Use digital health solutions for the new model of care (state solutions for above).
- Support entrepreneurship and start-ups for health technology through research centers.
- After achieving the digital transformation in healthcare, shift into a data-driven strategy in healthcare.
- Include XR in remote healthcare monitoring.
- Promote machine diagnoses to minimize errors (Fig. 3).

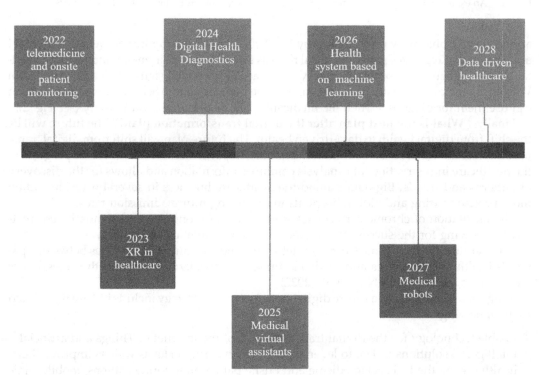

FIG. 3   Initiatives for a future digital transformation strategy.

2. Literature review

Digital technology has become an essential aspect of everyday life. Adopting digital healthcare services is not a luxury; it is a cornerstone of efficient health systems. A rapid application of digital technology in health sectors has been noticed during the COVID-19 pandemic. Moreover, the emerging deployment of remote healthcare services via telehealth and telemedicine during COVID-19 pandemic has resulted in shifting care delivery to be technology based.

Pervasive examples of these technologies are blockchain and AI. Blockchain has been implemented in the healthcare sector to combat the pandemic through data management, safe sharing, and meaningful utilization for tracking outbreaks (Negro-Calduch et al., 2021; Jabarulla and Lee, 2021). Additionally, AI technologies have offered machine-based solutions that analyze patient medical data to suggest appropriate treatment and predict future pandemics (Negro-Calduch et al., 2021).

During COVID-19, the interaction between the medical team and cancer patients was limited as those patients are at higher risk of infection and mortality. Hence, telehealth can allow healthcare providers to remotely monitor patient medical status, chemotherapy care, and palliative care (Paterson et al., 2020).

Digital health has played a major role in delivering healthcare services during COVID-19 in Saudi Arabia. Plenty telemedicine mobile applications were developed and their usability was evaluated in the literature (Aldekhyyel et al., 2021; Al-Hazmi et al., 2021), including:

- **Seha** is a free consultation application linked to the MoH and available to both citizens and residents.
- **Cura** is a paid consultation application.
- **Dr. Sulaiman Alhabib** is an application developed by the Dr. Sulaiman Alhabib private hospital to offer many medical services to patients, including remote consultations.

Other applications named Sehaty and Mawaid were used during the pandemic to check the health condition of suspected COVID-19 patients, where they can report their data as a form of surveillance. They can also be medically advised and followed up by specialists via the Tatamman application for remote care and enhancement of the recovery process (Al-Hazmi et al., 2021).

The use of digital technology in the healthcare system has been continuously expanding locally and worldwide. Global pandemics like COVID-19 shed light on the need for adopting technology-based solutions in the delivery of healthcare services.

3. A roadmap for the strategic transition

3.1 Digital transformation as an organizational developmental process

Among its strategic goals for achieving national transformation, Saudi Arabia aims to ease access to healthcare services, enhance their quality, and prevent health risks. The need for transformation arose from identifying major challenges that have been faced in the local healthcare system. These challenges are as follows: the continuously increasing population,

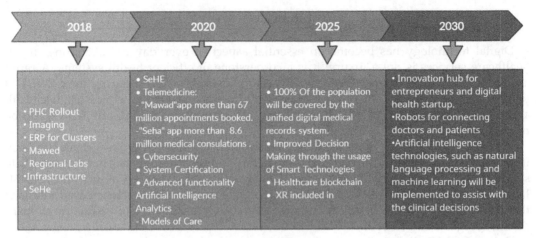

FIG. 4   The roadmap for the Saudi strategic e-Health transition.

high rates of preventable injuries, inadequate primary care, insufficient quality of healthcare services, resource and staff-centric system, workforce capacity, and patient economic status. Thus, many strategies have been developed to drive the transformation of healthcare considering the need for adapting digitalization and emerging technologies in the system (National Transformation Strategy, n.d.) (Fig. 4).

While the leader of this process is the MoH, other institutions are involved such as the Saudi Health Council, the Saudi Food and Drug Authority, King Faisal Specialist Hospital and Research Center, the Saudi Red Crescent Authority, and the Ministry of Education. National health transformation relies mainly on the utilization of e-health to increase healthcare service coverage and improve patient experience and system sustainability. Consequently, one of the national initiatives is e-health, which aims at increasing the efficiency of the Saudi healthcare sector by utilizing technology and implementing digital transformation. It works in a close collaboration with another initiative called the National Health Information Center-Cross Cutting e-Health Initiatives. Its main focus is to construct a more coherent healthcare system and integrate healthcare services to ease access to them. Establishing an electronic tracking system is another initiative for e-tracking distribution and the availability of pharmaceuticals. To improve public health and prevent epidemics, the HESN program (https://www.moh.gov.sa/en/Ministry/Projects/pre/Pages/News-2019-02-06-004.aspx) was introduced. Finally, the National Horizontal Health Records initiative was designed to target collecting data on the procedural management of predetermined diseases among the local population (National Transformation Program, 2018).

Health system governance has been one of the pillars of the Saudi healthcare transformation strategy. It is a system of policies aimed at regulating and monitoring performance in providing healthcare services to the population. This framework helps recognize requirements for improving data and digitization, showing the need for empowering the National Health Information Center, setting standards for health data interoperability, and pointing out the financial and clinical requisites of digital transformation. At the national level, investment in e-health is expected to result in several benefits as follows:

- **Improving patient experience,** as it eases access to healthcare services anytime and anywhere.
- **Increasing the safety and efficiency of services** by providing organized and accurate data, which can reduce medical errors and their consequential burdens.
- **Assisting healthcare professionals** in the decision-making process through automation of main administration-related tasks.
- **Sustaining an efficient health system** by integrating the databases and systems of different care settings in Saudi Arabia.
- **Creating new job and investment opportunities** for both the workforce and entrepreneurs, which benefits the Saudi health system and economy.

Unifying health records is a main target in the National Transformation Program 2020. To implement e-health in health records, the Saudi e-health strategy introduced an integration mechanism called the Saudi Health Exchange (SeHE) and accordingly suggested three classes of solutions:

- **National solutions** provided by one party to all system segments.
- **Cluster solutions** vary from one cluster to another.
- **Enterprise solutions** provided by many vendors to all system segments (Fig. 5).

The Saudi Data and AI Authority (SDAIA) is the governing body for AI solutions and initiatives. In 2019, SDAIA's transformation strategy was approved. It aims to drive the national data and AI and supervise their execution by setting policies, analyzing data, and encouraging innovation. An example of SDAIA's outcomes is the Tawakkalna application. This application was the fruit of a collaboration between the MoH and SDAIA to control COVID-19. Tawakkalna was developed primarily to electronically permit movement in and out of governmental and private institutions (SDAIA, n.d.). It has many special and useful features that contribute to curbing virus transmission by automating transactions between relevant parties.

FIG. 5   The interrelationship of solution classes and SeHE.

Even after citizens have returned to a normal life, Tawakkalna has been still used to facilitate everyday life tasks, such as governmental transactions for custody, financial affairs, and official documents. The health-related services provided by Tawakkalna during the curfew included requesting permission for movement electronically under specific circumstances such as for necessary supplies, work in institutions excluded from the curfew, and cases needing prompt medical care (Tawakkalna, n.d. https://ta.sdaia.gov.sa/en/index).

Some health-related services provided by Tawakkalna during the "cautiously we return" period are as follows:

- **COVID-19 vaccine**—for booking an appointment for COVID-19 vaccination after ensuring eligibility.
- **Health passport**—to show the status of the user's immunity against COVID-19, the number of vaccine doses they have taken.
- **Digital identities**—to preview the user's official documents such as national identification.
- **Previewing medical, civil affairs, and passport appointments.**
- **Providing care to dependents**—to allow custodians to request taking care of specific dependents so that they can view their information on the application.
- **Sharing health condition cards**—to allow custodians to share their dependents' health condition cards with other users.
- **COVID-19 test**—for booking an appointment for COVID-19 screening.
- **Health condition**—shows the health condition of users according to MoH encrypted data (Tawakkalna, n.d.).

Transition into e-health is a collaboratively organizational process. All relevant parties in Saudi Arabia work to incorporate AI and other technologies in the healthcare system to provide high-quality services to the local population. This approach has been adapted recently with the emergence of the national transformation strategy, and more applications and implementations of e-health are expected to be seen as time goes on.

3.2 Action plan and milestones for digital transformation in the Saudi National Institute of Health (SNIH)

The SNIH is one of the initiatives launched by the MoH in the national transformation of 2020 that aims to transform the country into a knowledge-based economy. SNIH supports health researchers and innovators. It provides innovation management that include all health-related services, including the development of medical devices and engineering. It also includes business models, design and manufacturing solutions, and various applications related to smart health care. Knowledge translation works to transform research and innovations into products and services through proposing policies, rules, and procedures for the application and implementation of knowledge. SNIH can be a resource not only for researchers and innovators but also serve as an innovation hub that attracts young entrepreneurs to transform the country into a hub ecosystem.

Moreover, it also includes building and investing in ideas as well as providing market and finance access to tech entrepreneurs. An innovation hub brings together various stakeholders in the innovation ecosystem in one location to share common goals and possibilities. The innovation hub focuses on researchers, inventors, and entrepreneur-driven economic growth through providing space, programming, and hosting events. The group raises awareness of entrepreneur-led innovation in the community by utilizing those three components.

This can happen with the support of the government and healthcare governmental sectors as well as partnerships with technology companies.

SNIH can be a major resource for digital transformation in healthcare, where it can be an organization that nurtures new ideas and start-ups for digital health solutions for ambitious innovators.

Two ways suggested for the innovation hub are:

1. Entrepreneurs and start-ups have an idea that can be executed with the support of SNIH.
2. Individual researchers, inventors, and entrepreneurs come together to build a start-up with the support of SNIH.

4. Conclusions

Saudi Arabia's digital transformation began in 2018, with the implementation of Vision 2030's rapid digital change. SeHE, telemedicine, cybersecurity, system certification, advanced functionality, artificial intelligence, analytics, blockchain, and care models are among the systems under consideration. The mission of the 2020 digital transformation plan was updated to "transform healthcare delivery through technology to deliver safer, more efficient healthcare services to the Saudi population" and encouraged the use of digital health.

During the COVID-19 pandemic in Saudi Arabia in 2020, a significant digital response emerged. Telemedicine is now available, as are telehealth apps such as Sehaty, Tawaklna, and Tataman. The MoH's mission has shifted to prevention, with a focus on noncommunicable diseases.

This chapter outlined the requirements for the transition to e-health in the age of artificial intelligence by identifying the major challenges and discussing the issues confronting the Saudi health system, such as data, electronic records, and clinical issues. Then, using digital health solutions and discussing the future of healthcare with new technology, we must implement an action plan that includes suggestions for future digital health transformation. All these transitions and issues for digital transformation will then be discussed in depth as an organizational developmental process by utilizing the transformation strategy, amplifying national security provided by SeHE, and cluster and enterprise solutions. Finally, the action plan and milestones for digital transformation in SNIH will be discussed, with the goal of transforming SNIH into an innovation hub that attracts business.

References

Aldekhyyel, R.N., Almulhem, J.A., Binkheder, S., 2021. Usability of telemedicine mobile applications during COVID-19 in Saudi Arabia: a heuristic evaluation of patient user interfaces. Healthcare (Switzerland), 9. https://doi.org/10.3390/healthcare9111574.
Al-Hazmi, A.M., Sheerah, H.A., Arafa, A., 2021. Perspectives on telemedicine during the era of covid-19; what can Saudi Arabia do? Int. J. Environ. Res. Public Health, 18. https://doi.org/10.3390/ijerph182010617.

Almalki, M., Jamal, A., Househ, M., Alhefzi, M., 2021. A multi-perspective approach to developing the Saudi Health Informatics Competency Framework. Int. J. Med. Inform., 146. https://doi.org/10.1016/j.ijmedinf.2020.104362.

Jabarulla, M.Y., Lee, H.-N., 2021. A blockchain and artificial intelligence-based, patient-centric healthcare system for combating the COVID-19 pandemic: opportunities and applications. Healthcare 9 (8), 1019. https://doi.org/10.3390/healthcare9081019.

National Transformation Program. 2018.

National Transformation Strategy. n.d.

Negro-Calduch, E., Azzopardi-Muscat, N., Krishnamurthy, R.S., Novillo-Ortiz, D., 2021. Technological progress in electronic health record system optimization: systematic review of systematic literature reviews. Int. J. Med. Inform., 152. https://doi.org/10.1016/j.ijmedinf.2021.104507.

Nico, M., Garante, D., Valtorta, K., Kharbanda, V., Sehlstedt, D., 2022. Data-Driven Healthcare|Arthur D. Little (online). Adlittle.com. Available at: https://www.adlittle.com/en/insights/viewpoints/data-driven-healthcare. (Accessed January 2021).

Paterson, C., Bacon, R., Dwyer, R., Morrison, K.S., Toohey, K., O'Dea, A., et al., 2020. The role of telehealth during the COVID-19 pandemic across the interdisciplinary cancer team: implications for practice. Semin. Oncol. Nurs., 36. https://doi.org/10.1016/j.soncn.2020.151090.

SDAIA n.d. https://sdaia.gov.sa/?Lang=en&page=SectionAbout# (Accessed 24 February 2022).

Tawakkalna n.d. https://ta.sdaia.gov.sa/en/index (Accessed 24 February 2022).

Wilson, A., Saeed, H., Pringle, C., Eleftheriou, I., Bromiley, P.A., Brass, A., 2021. Artificial intelligence projects in healthcare: 10 practical tips for success in a clinical environment. BMJ Health Care Inform. 28, e100323. https://doi.org/10.1136/bmjhci-2021-100323.

A bibliometric analysis of GCC healthcare digital transformation

Tayeb Brahimi[a] and Akila Sarirete[b]

[a]Energy and Technology Research Lab, College of Engineering, Effat University, Jeddah, Saudi Arabia [b]Computer Science Department, College of Engineering, Effat University, Jeddah, Saudi Arabia

1. Introduction

Artificial Intelligence (AI) has been identified as one of the most important facilitators of digital transformation in several industries, including healthcare (Holmstrom, 2021; Ivančić et al., 2020). Since the outbreak of the COVID-19 pandemic, global research has made tremendous progress in bringing together the world's scientists and international health experts. Scientists used digital technology to fight the pandemic, contain its rapid transmission, discover therapeutic measures, accelerate the development of vaccines, and become more prepared for the future. Two critical strategic response and preparedness plans (SPRP20 and SPRP21) were launched by the World Health Organization (WHO) in February 2020 and 2021 (WHO, 2020, 2021), in addition to the Access to COVID-19 Tools (ACT) Accelerator launched at the end of April 2020 (WHO, 2022). The SPRP20 strategic plan stresses the progress and achievement made in international coordination, support, country preparedness, research, and innovative technologies; the SPRP21 plan focuses on overcoming the ongoing challenges in response to COVID-19, including risks related to vaccines and the emergence of SARS-CoV-2 variants of concern (VOC). On June 11, 2021, UNESCO released the Science Report "The Race Against Time for Smarter Development" (Schneegans et al., 2022), which calls for significant scientific investment to address growing crises and achieve a digitally and environmentally smart future. According to ResearchGate's survey (ResearchGate, 2020), nearly half the researchers surveyed spent more time reading, studying, and analyzing scientific medical work and research than before the pandemic. The same pattern was also found in the time spent writing

AI-Covid Applications

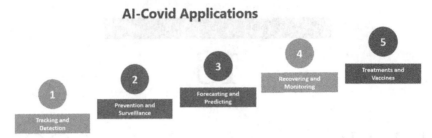

FIG. 1   AI applications at different stages of the COVID-19 crisis.

and publishing peer-reviewed scientific articles before and after the pandemic. Researchers have great optimism that AI can be used as a tool and driving force in diagnosing, tracking, and forecasting COVID-19 outbreaks (Krit et al., 2022). Fig. 1 shows a nonexhaustive list of some AI applications at different stages of the COVID-19 crisis. Recently, AI has played a key role in developing models to study virus transmission, identify high-risk patients, and control the pandemic's spread (Jiang et al., 2020; Nguyen et al., 2021). For example, a call to action for AI experts worldwide to develop AI tools and help scientists in the fight against the pandemic was launched in March 2020 by the White House Office of Science and Technology Policy (HealthITAnalytics, 2020; Huergo, 2020). Based on existing publications, a report published in May 2021 by the American Association for the Advancement of Science (AAAS) (Kimbrell, 2020) highlighted several ways in which AI supported the pandemic. This report suggested grouping AI applications into five broad classes: forecasting, diagnosis, containment and monitoring, drug development and treatments, and social and medical management. Also, the WHO published a health report on AI suggesting six important principles for equitable, safe, sustainable, and maximum benefit (Jasarevic, 2020): protecting autonomy; promoting human well-being, safety, and the public interest; ensuring transparency, explainability, and intelligibility; fostering responsibility and accountability; ensuring inclusiveness and equity; and promoting AI that is responsive and sustainable.

Due to the increasing number of scientific publications on COVID-19 and AI, bibliometric analyses, data visualization, and mapping have been used successfully to study statistical behavior and assess scientific activity in various technology and healthcare fields, including COVID-19 and AI (Alcalá-Albert and Parra-González, 2021; de Las Heras-Rosas et al., 2021; Gao et al., 2021; know.space, 2021; Nobanee et al., 2021). In its most basic definition, bibliometrics applies quantitative analysis and statistics to academic publications such as journals, articles, and books and the number of citations they received (Andres, 2009). Many governments have created modern bibliometrics programs with large teams of analysts that regularly release bibliometric reports, commonly referred to as "scientific indicators studies" (Pendlebury, 2008). The popularity of bibliometric analysis has been facilitated by the development of several tools such as CiteSpace and VOSviewer (Moral-Muñoz et al., 2020), and databases such as the Web of Science (WoS), Scopus, the Lens, or Google Scholar (Pranckutė, 2021). In this study, VOSviewer (van Eck and Waltman, 2009) has been selected because it is particularly well suited for viewing large networks and uses distance-based visualizations instead of graph-based ones. The distance between two nodes reveals how the nodes are linked. To the best of the author's knowledge, the present study is the first bibliometric study to analyze thoroughly COVID-19 and AI within the GCC region. The present study seeks to assess the various contributions of the six countries of the Gulf Cooperation Council (GCC) (Saudi Arabia, Qatar, the United Arab Emirates (UAE), Kuwait, Oman, Bahrain), in terms of scientific publications on AI and COVID-19, with particular emphasis on digital transformation and the methods adopted to fight the COVID-19 pandemic and protect community well-being. The rest of the chapter is organized as follows: an overview of the state-of-the-art bibliometric analysis is presented in the next section. Section 3 describes the research methodology for generating various bibliometric visualizations and trends. Section 4 covers the research output, relationship, and mapping among different scientific publications in the GCC region, followed by a discussion of the results. Section 5 presents an example of a learning activity using WoS and the Lens databases. Finally, Section 6 presents the conclusion and future work of the present study.

2. Literature review

According to the 2022 Artificial Intelligence Index Report (Zhang et al., 2022), in 2021 the number of AI publications in China was 63.2% higher than in the United States. The number of patents filed was 30 times higher than that in 2015. The total number of AI publications increased from 162,444 in 2010 to 334,497 in 2021, with 51.5% as journal articles and 21.5% as conference papers. In the Scopus database, the number of coronavirus AI-related publications in 2020 and 2021 was dominated by medicine (21%) and computer science (11%) (Aljahdali et al., 2021). Due to COVID-19, AI conferences took place virtually, which increased the number of attendees and the number of papers presented. In 2020, for example, the number of AI publications increased by 34.5% compared to 2019 (Zhang et al., 2021). The highest number of these publications originated from academic institutions globally, with China surpassing the United States in citations (Zhang et al., 2022). As the number of joined efforts to tackle the coronavirus continues to grow, so does the scientific community's number of scientific publications and reports. Since the outbreak of the COVID-19 pandemic in December 2019, a huge amount of COVID-19-related resources have been published in journals or as a preprint. This is in addition to the availability of automated search engine tools used as comprehensive resources on COVID-19 or for asking complex scientific questions on the disease such as COVID Scholar, LitCOVID, WHO database, iSearch COVID-19 portfolio, or the Centers for Disease Control and Prevention (CDC), to cite only a few (COVID-19 Map, 2022; Stephen, 2020). A simple search using "COVID-19" as the query for the period 2020–21 in Google Scholar or Scopus returned more than 100,000 documents. Assessing such a volume and impact of scientific research is not an easy task (Palayew et al., 2020); To keep current with these fast-evolving efforts and extract information from a large number of research papers, we are conducting comprehensive data mining of scientific literature records to discover development trends, research hotspots, most active research institutions and countries, keywords, and citation trends on COVID-19 and AI. Researchers have used different approaches to qualitatively and quantitatively analyze past research findings using publication metrics. Vaishya et al. (2020), in their paper "Artificial Intelligence applications for COVID-19 pandemic," identified seven AI applications for COVID-19 that include early detection and diagnosis, treatment and monitoring, tracing of infected patients, vaccines and drugs, healthcare workload management, disease prevention, and real-time data analysis for better decisions. Kumar et al. (2020) examined the various aspects of the role of modern technologies to combat the COVID-19 pandemic, including how AI makes it possible to identify, track, and forecast outbreaks. However, the authors concluded that only a few modern technologies have been sufficiently developed. Adadi et al. (2021) investigated research on AI in the post-COVID-19 phase. They found that the number of works discussing the time after COVID-19 has been limited so far. In another study on the application of digital health to COVID-19, Gunasekeran et al. (2021) investigated the application of AI in the first 6 months of the appearance of COVID-19. While the authors recognized the significant progress and promised to alleviate the consequence of COVID-19, they also reported that some of these digital solutions have limitations, mainly due to the constraint of clinical trials and operational applications. According to health officials and the WHO (Feuer and Higgins-Dunn, 2020), one of the big challenges is to understand "asymptomatic patients, "i.e., finding the reason why some persons who are infected with the coronavirus do not get sick. In this context, the MIT AutoID Laboratory and Harvard University (Laguarta et al., 2020) developed an AI screening tool trained to detect COVID-19 asymptomatic patients using only cough recordings. The MIT Open Voice model achieved promising results in increasing approaches used to contain the spread of the pandemic. The model reached COVID-19 sensitivity of 98.5%.

As we can see, there is a growing number of scientific literature publications and many researchers are now turning to bibliometric analysis as a powerful tool for discovering, identifying, visualizing, and evaluating scientific research by applying quantitative techniques to bibliometric data (Donthu et al., 2021) such as analyzing the AI and COVID-19 body of knowledge and trends from published literature. The term "bibliometric" appeared in 1969 and was defined as the "application of mathematical and statistical methods to books and other media of communication" (Pritchard, 1969). The primary goal of bibliometrics is to evaluate the scientific literature on a specific topic and assess the quality of individual intellectual contributions. A large number of papers examined the definition of bibliometrics as well as its growth into a useful research tool, such as the studies by Broadus (1987) and Hood and Wilson (2001). Other metric studies also appeared and are becoming popular such as scientometrics, webometrics, cybermetrics, informetrics, and altmetrics. The type of analysis includes coauthorship with the unit of analysis: documents, sources, authors, organizations, and countries; cooccurrence with the unit of analysis: all keywords, author keywords, and index keywords; citation with the unit of analysis: documents, sources, authors, organizations, and countries; bibliometric coupling with the unit of analysis: documents, sources, authors, organizations, and countries. Table 1 briefly defines each of the metric studies used and their corresponding references.

Hu et al. (2020) applied bibliometric visualization for medical data mining (MDM) to investigate the research topics and development trends in medicine. The authors found that the hot topics are drug discovery, medical imaging, and vaccine safety. The most active country was the United States, followed by China and India. Islam et al. (2021) conducted a systematic literature search on AI in COVID-19 using the WoS database between Feb. 1, 2020, and Feb. 1, 2021. In this analysis, the authors found China most active in publishing articles on AI and

TABLE 1 Metric studies for article analysis.

Metric	Definition	Reference
Bibliometrics	Application of mathematical and statistical methods to books and other media of communication	Pritchard (1969)
Scientometrics	Application of quantitative methods for the analysis of science is viewed as an information process	Wolfram (2003)
Webometrics/ cybermetrics	Study the quantitative aspects of scientific literature from electronic resources/websites such as hyperlinks, the structure of the website, etc.	Björneborn and Ingwersen (2001)
Informetrics	Study the quantitative aspects of information in any form, not just records or bibliographies, and in any social group, not just scientists	Almind and Ingwersen (1997)
Almetrics	Alternative metrics study and measure the impact of research through social networks	Sud and Thelwall (2014)

COVID-19, followed by the United States and India. In addition, China was the most active in collaborating with the United States, Italy, and Canada. A special issue on "Bibliometric Studies and Worldwide Research Trends on Global Health" published by the International Journal of Environmental Research and Public Health (Salmerón-Manzano and Manzano-Agugliaro, 2020) used bibliometric analysis to explore four areas of global health: research on diseases, health, environment, and education society. Most research studies on COVID-19 focused on public health, viral pathogenesis, and clinical management; however, psychosocial difficulties and the influence of COVID-19 on different vulnerable populations have received minimal consideration, which indicates a need for research collaboration between countries.

The GCC is rapidly evolving and undergoing drastic change due to AI technologies and digital transformation (Brahimi and Bensaid, 2019; Durou et al., 2017), particularly in countries such as the UAE and Saudi Arabia. In recent years, the GCC governments have implemented national plans for digital transformation and embarked on a series of massive transformations toward the Fourth Industrial Revolution characterized by digital transformation, innovative technologies, knowledge-based economy, information society, e-government, smart cities, smart communities, smart tourism, and next-generation care. The GCC states' ambitions are based on their respective visions and investment plans; for example, Saudi Arabia established its Vision 2030 on a vibrant society, thriving economy, and ambitious nation. The UAE Vision 2021 focuses on world-class healthcare, a competitive knowledge economy, a safe public, and a fair judiciary. The Qatar Vision 2030 pillars are economic development, social development, human development, and environmental development. The Kuwait Vision 2035 is set to move the country toward being a financial and trade hub with a sustainable, diverse economy as well as a high-quality living environment and healthcare. Oman's theme for Vision 2040 is a society of creative individuals, a comprehensive economy, and a state with responsible apparatus. In this strategic plan, health has a high priority with the ambition of covering all the regions of Oman with a comprehensive and fair healthcare system. The Bahrain Vision 2030 aspires to move from an economy based on oil to a productive, globally competitive economy using three guiding principles: sustainability, competitiveness, and fairness, with equal access to education, healthcare, and the job market.

From each country's vision, healthcare is considered one of the top priorities. Coupled with the new digital transformation in the region, healthcare has led to several digital initiatives to develop world-class public health. The initiatives include implementing a complete national e-health system, engaging in health tech innovation, online preventive and digital awareness health systems, and virtual and mobile health services. As shown in Table 2, each program includes a restructuring of the health sector (Durou et al., 2017; GCC Stat, 2011). For example, in Saudi Arabia (KSA, 2022), under the health sector, we can find "Improve Healthcare Service," "Promote a Healthy Lifestyle," "Improve Livability in Saudi Cities," "Ensure Environmental Sustainability," "Promote Culture and Entertainment," and "Create an Empowering Environment for Saudis." The key objectives of the Saudi health sector transformation program are: (1) Facilitating access to healthcare services, (2) improving the quality and efficiency of health services, (3) promoting prevention of health risks, and (4) enhancing traffic safety. It is expected that by 2025 the unified digital medical records will cover 100% of the Saudi population.

TABLE 2   GCC national visions and plans.

Country	Transformation program	National plan and healthcare key objectives	Examples of digital initiatives
Bahrain		*Bahrain National Planning Development Strategy.* Providing world-class and sustainable healthcare of modern medicine by improving the country's health system, including a healthy lifestyle, equitable access, and high-performance ethic	Digital Government Strategy 2022, health eServices, digital-first principle public services, Commercial Registration Portal (Sijilat)
Oman		*Oman National Program.* Establishing Oman as a regional and global leader in high-quality health research. Released the Health Vision 2050 plan for health research	Digital Government Strategy 2022, e. Oman, e.Government, digital society, Oman's Nebula Artificial Intelligence platform
Qatar		*National Health Strategy.* Providing a comprehensive, integrated, and preventive world-class healthcare system with effective and affordable services and high-quality research	Smart Qatar or TASMU, Searchable City, Digital Travel Guide, Augmented City, and Contextual Indoor Navigation
Kuwait		While restoring the regional leadership role of Kuwait as a financial and commercial hub in the health sector, the country aims at improving the quality and efficiency of its services. It works hand in hand with the private sector	The new Kuwait, Internet of Things, e-commerce, online shopping, and B2B telecom solutions

TABLE 2 GCC national visions and plans—cont'd

Country	Transformation program	National plan and healthcare key objectives	Examples of digital initiatives
Saudi Arabia	VISION 2030 رؤية KINGDOM OF SAUDI ARABIA المملكة العربية السعودية	*National Transformation Program.* Facilitating access to healthcare services, improving the quality and efficiency of health services, promoting the prevention of health risks, and enhancing traffic safety	Saudi Data and Artificial Intelligence Authority (SDAIA), Tech Programs (Tuwaiq, Hemmah, Qemmah), and Rashaka health awareness campaign
United Arab Emirates	رؤية VISION 2021 UNITED ARAB EMIRATES	*National Agenda.* Providing world-class healthcare, collaborating with all health authorities, improving the healthcare system, reducing lifestyle-related diseases, and excelling in preventive medicine	Dubai Internet City, UAE Artificial Intelligence (AI) Strategy, Emirates Blockchain Strategy, Middle Eastern ridesharing

In comparison to the top COVID-19 hit countries such as the United States, India, Brazil, the United Kingdom, Russia, and France, the GCC countries quickly took all necessary measures to limit the spread of the pandemic by imposing inward and outward travel bans, country lockdowns, closures of schools and universities, bans on mass gatherings, social distancing, and self-isolation. Other restrictions were also imposed in Saudi Arabia (Yezli and Khan, 2020), such as suspending Umrah (the voluntary Muslim pilgrimage to Makkah), daily and Friday prayers in about 80,000 mosques, and public transportation. Since the pandemic's beginning, the GCC countries have carefully monitored the COVID-19 situation, including healthcare, education, economy, and business, to protect their community's well-being while providing free medical care to all COVID-19 patients.

The GCC countries performed outstanding work in limiting the massive spread of the COVID-19 pandemic by implementing proactive strategies and making prompt choices in response to the COVID-19 pandemic. Raimondo et al. (2022) reported that with the increased interest in AI and health, the GCC healthcare sector shows that more than 50% of the GCC strategies include AI and data analytics to reduce costs and increase efficiency in the healthcare sector. For example, the department of health in the UAE developed the TraceCOVID App to detect and trace people who have been in close contact with those diagnosed with COVID-19. Also, Dubai police deployed an Oyoon monitoring system to enforce the lockdown. In Bahrain, the App BeAware tracks proximity to an individual with COVID-19 based on location data. Saudi Arabia has launched several apps to help combat the pandemic and provide various health services to its citizens, including Tetamman (rest assured) to reinforce isolation commitment, Tabaud (social distancing) to track the spread of COVID-19, Tawakkalna (COVID-19 KSA) to prevent the spread of COVID-19, and Sehhaty to provide health services for citizens (Hidayat-ur-Rehman et al., 2021). Table 3 shows the leading apps for COVID-19 used in the GCC.

TABLE 3   List of COVID-19 apps in the GCC.

Country	Application	Role	Reference
Bahrain	BeAware	A contact tracing process to identify and keep track of COVID-19 active cases and their contacts	MOH (2022)
Kuwait	Shlonik	An application that provides protection and treatment to those placed in isolation or quarantine	CAIT (2021)
Oman	Tarassud	Provides information on the current disease, access to medical hotlines, and acts as a surveillance system	MOH (2021)
Qatar	Ehteraz	Notifies when a citizen has been exposed to a person who is suspected of having an infection or who has been confined	ACTA (2021)
Saudi Arabia	Tawakkalna	Tracks coronavirus and shows user health status. It includes a variety of other healthcare-related services	SDAIA (2021)
Emirates	Alhosn	A contact tracing process to control infectious disease outbreaks and keep track of COVID-19	UAE (2022)

3. Research methodology

The quantitative methodology used in this study to generate various bibliometric visualizations and trends is based on retrieving a set of publications from the Scopus, WoS, and lens. org databases. This study extracted various data types from the three databases (Pranckutė, 2021). Then we used quantitative techniques to generate multiple bibliometric visualizations and trends based on the retrieved data. Scopus and WoS are the two bibliographic databases often regarded as the most comprehensive data sources for multiple disciplines (Pranckutė, 2021). The Lens database (Lens, 2022) is a free platform of a global open cyberinfrastructure serving more than 200 million scholarly records combining Microsoft Academic, PubMed, and Crossref, enhanced with UnPaywall open-access information and Patent datasets. The systematic search was performed on Dec. 31, 2021, using the three databases for articles published in the last 5 years from January 2017 to December 2021. However, in this study, the focus is on the results from the Scopus database, as WoS and Lens are left as assignments as described in Section 5. The literature analysis was based on citations, cocitations, authors, journals, organizations, countries, keywords, etc., using the software package VOSviewer version 1.6.17, released on July 22, 2021 (van Eck and Waltman, 2022). The selection of the databases follows the guidelines as published as discussed by Andres (Andres, 2009), namely: (i) the coverage by the database, (ii) the consistency and accuracy of the data, (iii) the data fields required by the study, (iv) the available browsing and search options, (v) the availability of built-in analytical tools for preliminary exploration and analysis, and (vi) the availability of saving and exporting options.

The focus of the output concerns the productivity of the scientific publication, the topic of study, the collaboration and connectivity among countries, the corresponding strength, the most used keywords, the type of scientific journals, and the document or country citations and cocitation links. This study presents publication patterns of the top most productive countries, published topic areas, and most productive keywords, among other previously described characteristics. The collected data were restricted to the six countries of the GCC. VOSviewer displays the output in three ways: (i) network visualization, (ii) overlay visualization, and (iii) density visualization. The color of a point on a map is determined by the density of items in the area. For example, the objects (publications, keywords, etc.) are represented by their labels and circles in the network visualization. The label and circle of an item increase in size with weight. The color of an element is determined by its cluster while the lines represent the corresponding links. Less distance between two items means more relatedness. Finally, each link possesses a degree of strength, represented numerically by a positive value. The greater the value of this variable, the stronger the relationship is. Note that VOSviewer version 1.6.17 can create maps that contain many thousands of objects, and it is also capable of displaying maps with more than 10,000 objects.

Lens.org database can retrieve up to 50,000 items while Scopus is limited to 2000 items and WoS is limited to 500 items per execution. The process of creating a map is broken down into three phases. A similarity matrix is first constructed using a cooccurrence matrix. The VOS mapping approach is applied to the similarity matrix to create a map in the next step. The map is translated, rotated, and reflected (van Eck and Waltman, 2009). The VOS mapping approach creates a two-dimensional map in which the items 1, ..., n are placed so that the distance between any two items i and j represents their similarity. The main steps of the bibliometric analysis used in this study are given in Fig. 2.

In the next sections, a step-by-step method for extracting documents is given for the three databases. However, in the present study, we will focus on the result from the Scopus database only; WoS and the Lens will be left for the reader as practical examples. For details about each database, see references Pranckutė (2021), Falagas et al. (2008), and Lens (2022) to get more in-depth information on how to use these databases.

3.1 Scopus database

Scopus is a commercial and subscription-based product with a high price that Elsevier launched as a comprehensive bibliographic data source in 2004 (Pranckutė, 2021). Since then, Scopus has earned equal standing or better than the WoS database. According to the Scopus website (Elsevier, 2022), the Scopus database contains up to 80 million documents, 234,000

Steps of the Bibliometric Analysis

Database Selection
Selecting the appropriate database

Search Criteria
Defining the search keywords

Search Results
Refining the search results

Initial Data
Screening and eligibility

Data Analysis
Visualization and mapping

FIG. 2 Main stages of the bibliometric analysis.

books, 7,000 publishers, and 80,000 institution profiles. To create a bibliometric analysis using the Scopus database, a user needs first to retrieve data from the database. Unfortunately, this is not free. The following steps are examples of retrieving published documents from Scopus (Beatty, 2016; Goodman, 2022):

(a) Create a Scopus account at https://www.scopus.com/ via your institution. Scopus will direct you to the steps needed for creating an account.
(b) Once the registration is completed, log on to Scopus. By default, Scopus opens the "Start Exploring" window with the search documents within the article title, abstract, and keywords.
(c) In the search box provided (see Fig. 3), type in the keywords and search terms you're looking for. For more details on the query string, click on the Advanced field, including operators and field codes.
(d) Use the "Limit to" or "Exclude" buttons to refine results.
(e) To analyze all the results click on "All," then click on the "Analyze Search Results" button.
(f) To export the result, click on "CSV export"' and save the file that will be uploaded later by VOSviewer software.

3.2 The WoS database

Like the Scopus database, WoS is a subscription-based access database providing citation data for 1.9 billion cited references from more than 171 million records. It is the first database

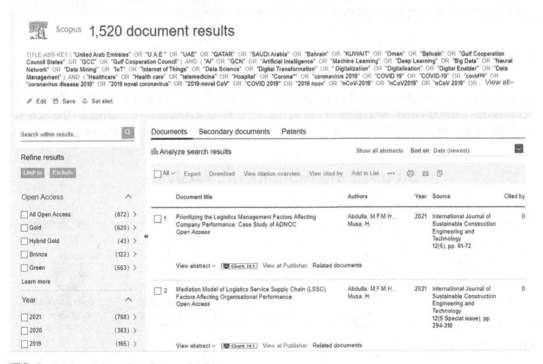

FIG. 3   Main window of the Scopus database.

and was founded in 1960; it was originally known as the Institute for Scientific Information (ISI) and is currently maintained by Clarivate (Pranckutė, 2021). WoS contains the Core Collection (WoS CC) with Science Citation Index Expanded (SCIE); Social Sciences Citation Index (SSCI); Arts & Humanities Citation Index (A&HCI); Conference Proceedings Citation Index (CPCI); Books Citation Index (BKCI); and Emerging Sources Citation Index (ESCI). The following steps are an example of how to retrieve published documents from WoS:

(a) Create a WoS account using https://www.webofscience.com/ via your institution.
(b) Once the registration is complete, log on to WoS.
(c) In the search field (see Fig. 4), type in the keywords and search terms you're looking for.
(d) To refine results, use operators and field codes.
(e) To analyze the results, click on the "Analyze" button.
(f) To export the result, click on "Export" and save the file that will be uploaded later by VOSviewer software. WoS allows downloading 500 records at a time; for more records, use the download in many batches.

3.3 The Lens database

Founded in 1998 as a Patent Lens, the Lens database is a free collection of databases that mainly includes more than 200 million scholarly records combining Microsoft Academic, PubMed, and Crossref, enhanced with UnPaywall open-access information and Patent datasets. Its primary goal was to access patent and scholarly documents for free and make them more transparent to all (Lens, 2022; Penfold, 2020). Because Lens is free, it makes it particularly attractive to healthcare libraries, which generally have limited resources compared to colleges and research organizations.

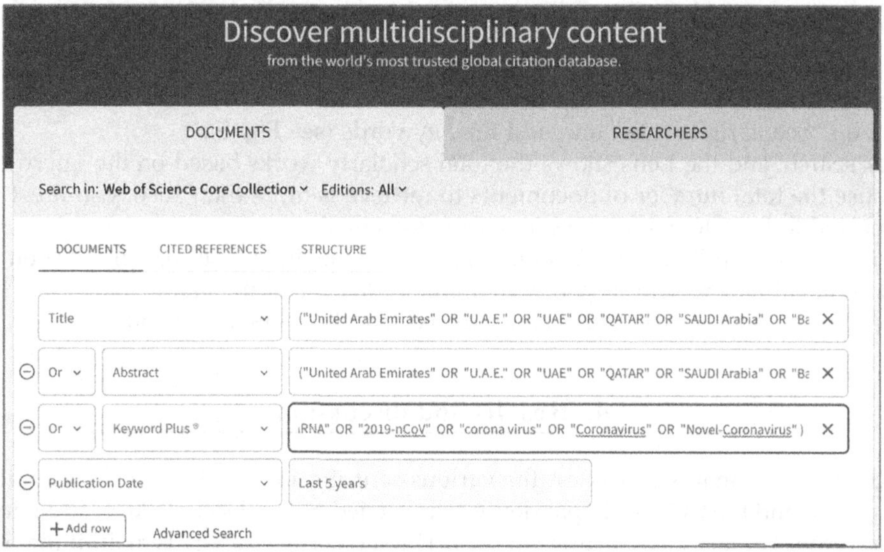

FIG. 4   Main window of the WoS database.

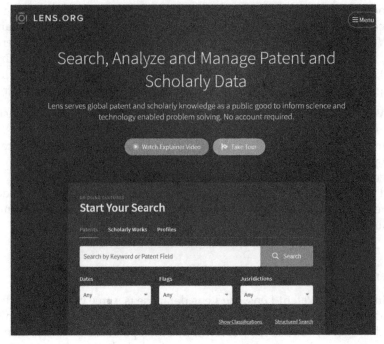

FIG. 5   Main window of the Lens database.

Recently, Lens has been added to VOSviewer version 1.7 (van Eck and Waltman, 2022) to support creating maps based on files exported from Lens. To make a bibliometric analysis, a user needs first to retrieve data from the Lens database (see Fig. 5). The following steps are an example of how to retrieve published documents from Lens.

(a) Go to the website https://www.lens.org/.
(b) Register for free, then log in to the website.
(c) Click on "Scholarly Works" and add the keywords (see Fig. 5).
(d) Click search, and the Lens shows the total scholarly works based on the query in (c).
(e) Because the total number of documents to retrieve is more than 1000, you must be connected to Lens to get all the documents via email.
(f) Click on Export and select the document number to include (use 10,000). Click on Export, and an email will be sent to you with a link to download the result.
(g) Save the file as CVS, which will be uploaded later by VOSviewer software.

4. Results and discussion

The bibliometric analysis to assess the various contributions of the GCC in scientific publications on AI and the COVID-19 pandemic was carried out on Dec. 31, 2021, and it included the period from January 2017 to January 2021. The screening strategy is to limit the search to the English language, exclude any duplication and include publications from 2017 to 2022.

The process uses the string TITLE-ABS-KEY: TITLE-ABS-KEY ("United Arab Emirates" OR "U.A.E." OR "UAE" OR "QATAR" OR "SAUDI Arabia" OR "Bahrain" OR "KUWAIT" OR "Oman" OR "Bahrain" OR "Gulf Cooperation Council States" OR "GCC" OR "Gulf Cooperation Council") AND ("AI" OR "Artificial Intelligence" OR "Machine Learning" OR "Deep Learning" OR "Big Data" OR "Neural Network" OR "Data Mining" OR "IoT" OR "Internet of Things" OR "Data Science" OR "Digital Transformation" OR " Digitalization" OR "Digitalisation" OR "Digital Enabler" OR "Data Management") AND ("Healthcare" OR "Health care" OR "telemedicine" OR "Hospital" OR "Corona*" OR "coronavirus 2019" OR "COVID 19" OR "COVID-19" OR "COVID19" OR "coronavirus disease 2019" OR "2019 novel coronavirus" OR "2019-novel CoV" OR "COVID 2019" OR "2019 ncov" OR "nCoV-2019" OR "nCoV2019" OR "nCoV 2019" OR "Severe acute respiratory syndrome coronavirus 2" OR "2019-ncov" OR "SARS-CoV-2" OR "SARSCov2" OR "SARS-CoV-2" OR "mRNA" OR "2019-nCoV" OR "corona virus" OR "Coronavirus" OR "Novel-Coronavirus") AND (LIMIT-TO (PUBYEAR, 2021) OR LIMIT-TO (PUBYEAR, 2020) OR LIMIT-TO (PUBYEAR, 2019) OR LIMIT-TO (PUBYEAR, 2018) OR LIMIT-TO (PUBYEAR, 2017)) AND (LIMIT-TO (LANGUAGE, "English")). The output is then verified by a manual assessment of the retrieved articles.

4.1 Overall publication trends

Using the above search string, we explored the publication of documents related to "bibliometrics" in the Scopus database. As discussed in previous sections, Fig. 3 shows the number of published papers about bibliometrics increasing in the last 20 years. Using a combined field that searches abstracts and document titles for the keyword "bibliometric" or "bibliometrics," the Scopus database retrieved 27,563 documents. As shown in Fig. 6, this number was only 201 documents in 2000, then 824 in 2010, and jumped to 4583 in 2021, a

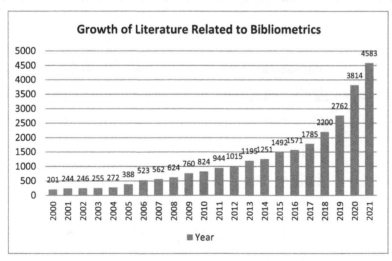

FIG. 6 Growth of Scopus-indexed publications related to bibliometrics from 2000 to 2021.

percentage increase of 456% compared to 2010. A similar process was done for the GCC region; in 2010, we retrieved four documents only; this number increased to 128 in 2021.

4.2 Publications trends in Scopus

Using the string described in Section 4, we explored the publication trends in AI, digital transformation, and COVID-19 retrieved from the Scopus database and limited the result to journals and conferences. The search identified 1520 hits. Removing duplicated and no author publications, we were left with 1479 documents. Fig. 7 shows the distribution of published papers by subject area. The most active country is Saudi Arabia with 711 records, then UAE with 251 documents, Qatar with 105 documents, Oman with 95, Kuwait with 68, and Bahrain with 60. The top-five most active organizations (see Fig. 8) are King Saud University (126 documents), King Abdulaziz University (104 documents), Imam Abdulrahman University (68), University of Sharjah (52 documents), and Qatar University (51 documents). As shown in Fig. 9, out of the 1479 documents, 20.6% were published in medicine, 18.6% in computer science, and 9.4% in engineering. The total number of published papers in the GCC in healthcare and AI increased from 107 in 2017 to 720 in 2021. Furthermore, the number of citations jumped from 44 in 2017 to more than 4600 in 2021, as shown in Fig. 10.

The network of cooccurrence of author keyword distribution, that is, keywords listed by the authors, is shown in Fig. 11. We use keywords provided by the paper's authors that have appeared more than five times in each retrieved document. In this case, 156 of the 4449 keywords fulfilled the requirement. The figure identified items in the network visualization using a circle. The cluster to which it belongs determines the color of an object and the lines connecting items represent the relationship among them. The greater the significance of an item, the bigger its label and circle. The size of the label and the size of the circle are determined by the weight of the object being labeled. According to Fig. 11 and based on the total number of documents and link strength, the top five dominant keyword occurrences were

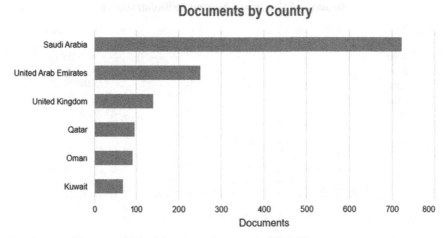

FIG. 7   Distribution of Scopus published documents by country (2017–21).

Documents by Affiliation

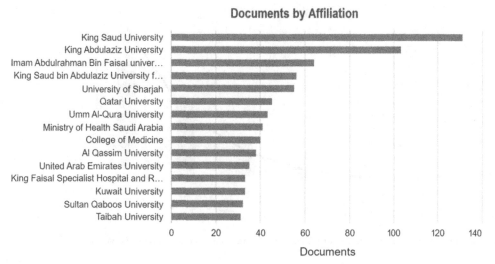

FIG. 8   Distribution of Scopus-published documents by affiliation.

Documents by Subject Area

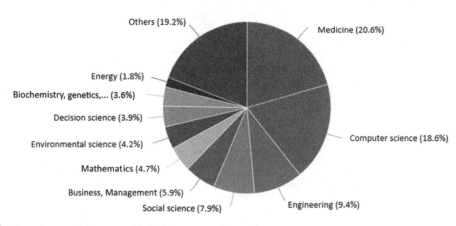

FIG. 9   Distribution of Scopus published documents by subject area (2017–21).

COVID-19 (247, TLS=410), Saudi Arabia (190, TLS=291), machine learning (62, TLS=121), artificial intelligence (44, TLS=79), and coronavirus (41, TLS=86) S=122). Fig. 12 shows the overlay visualization; unlike network visualization, elements in the overlay visualization are colored differently. The color bar shown in the top left corner of the visualization indicates the date of publication. The trend toward an increasing number of publications between 2019 and 2020 shows the importance of recent publications on COVID-19 and AI (yellow color; gray in print version). Moreover, Saudi Arabia is the most active country

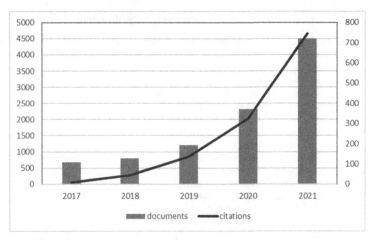

FIG. 10 Distribution of the number of published documents and citations over the years.

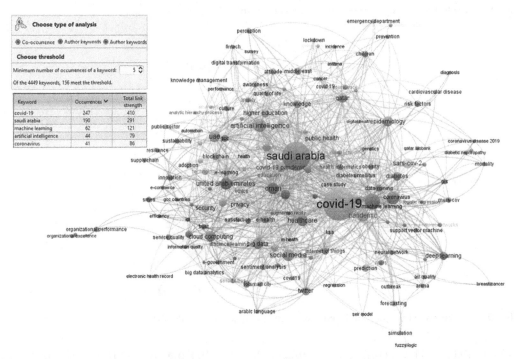

FIG. 11 Cooccurrence of author keywords. 4449 keywords, 156 meet the threshold (minimum of 5 occurrences).

among the GCC in the publication of articles on COVID-19 and AI, then comes Oman, UAE, Qatar, Bahrain, and Kuwait. A version of the item density visualization is shown in Fig. 13. In the same way as in network visualization and overlay visualization, the label of an item is used to show its item density visualization. In the item density visualization, each point has a color that shows how many items are at that point. By default, colors range from blue (dark

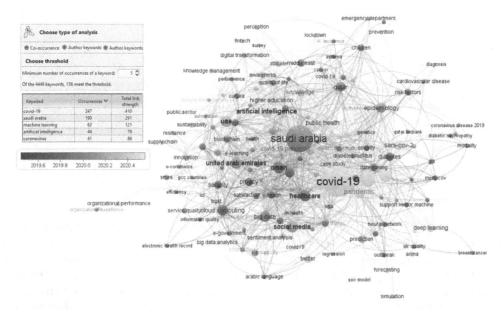

FIG. 12   Overlay visualization.

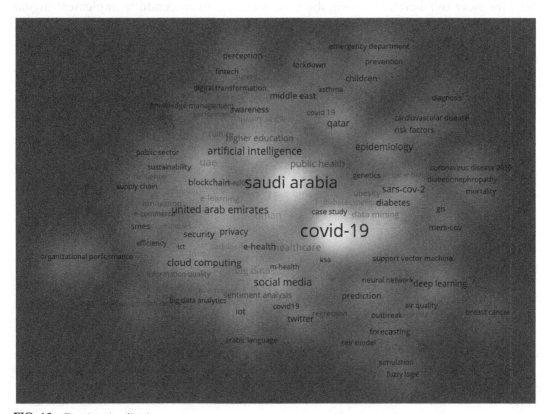

FIG. 13   Density visualization.

C. Global experiences and approaches to digital transformation in health

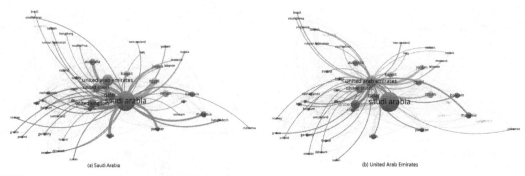

FIG. 14   Country collaboration: (A) Saudi Arabia and (B) UAE.

gray in print version) to green (light gray in print version) to yellow (gray in print version). The closer a point's color is to yellow (gray in print version), the more items are around it and the more weight those items have. On the other hand, the color of a point is closer to blue when there are fewer items around it and their weights are lower. In Fig. 14A (KSA) and B (UAE) as well as Fig. 15A (Oman), B (Bahrain), C (Qatar), and D (Kuwait), we attempted to display the activities and collaboration of each GCC country. The two GCC countries with the most collaborations are Saudi Arabia and UAE. The figure indicates a need for more collaboration among the GCC members to successfully implement digital

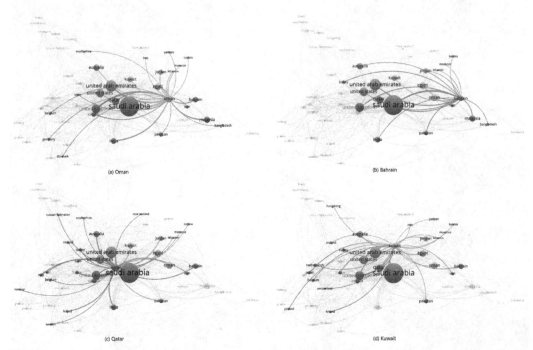

FIG. 15   Country collaboration: (A) Oman, (B) Bahrain, (C) Qatar, and (D) Kuwait.

C. Global experiences and approaches to digital transformation in health

health, effectively improve GCC primary healthcare, ensure security to citizens, and be well prepared for future pandemics.

5. Conclusion

In this chapter, we provided a comprehensive overview of AI-COVID-19-related research conducted in the Scopus, WoS, and Lens.org databases, including the various contributions of the GCC in terms of scientific publications to assist researchers, policymakers, and practitioners in better understanding the development of healthcare and AI research and the potential implications for practice. This study shows that the total number of COVID-19 and AI-related publications are dominated by medicine, computer science, social sciences, engineering, and biochemistry. Universities in Saudi Arabia and Qatar were among the top funding sponsors while Saudi Arabia and UAE were among the top in the number of publications. Saudi Arabia dominates the visualization network map of coauthorship. The study revealed that Saudi Arabia was the most active in terms of published documents, citations, and the number of organizations involved in research related to AI and COVID-19 or healthcare in general. However, it is also important to mention that the collaborations between GCC states need to be more aggressive. Bibliometric analysis shows that AI is on a steady upward trend, particularly in Saudi Arabia, Qatar, and the UAE. AI is becoming a game-changer in the GCC region. However, there is a long road ahead, which is why governments need to mobilize researchers and scientists in the GCC region and from all over the world to help fight the pandemic.

6. Learning activity: Bibliometric analysis

To achieve the attended learning outcome of discovering the power of bibliometric analysis, a range of activities has been prepared for the learner. In this activity, you will conduct a bibliometric analysis on suggested topics, or a topic of your interest, by first collecting publication data from different databases, then using VOSviewer software to visualize and analyze the output.

6.1 Activity description

Use the Scopus, WoS, or Lens databases, as explained in Section 4, to conduct a bibliometric analysis on one or more of the following topics and follow the steps described below:

(a) "Smart City"
(b) "Cybersecurity"
(c) Digital Transformation in Healthcare
(d) Virtual Reality and Augmented Reality
(e) Internet of Space
(f) Internet of Underwater Things or Underground Things

Step 1: Identify the objectives and scope of the bibliometric study.
Step 2: Select the techniques for conducting the bibliometric study.
Step 3: Compile the information for bibliometric evaluation.
Step 4. Run the bibliometric analysis and present your findings.

6.2 Recommended readings

1. Pranckute, R., 2021. Web of Science (WoS) and Scopus: The Titans of Bibliographic Information in Today's Academic World. Publ.
2. van Eck, N.J., Waltman, L., 2009. Vosviewer: A Computer Program for Bibliometric Mapping (SSRN Scholarly Paper No. ID 1346848). Social Science Research Network, Rochester, NY.
3. Penfold, R., 2020. Using the Lens Database for Staff Publications. Journal of the Medical Library Association : JMLA 108, 341–344.
4. van Eck, N.J., Waltman, L., 2022. VOSviewer—Visualizing scientific landscapes [WWW Document]. VOSviewer. URL https://www.vosviewer.com// (Accessed 1.10.22).
5. Elsevier, 2022. Scopus tutorials—Scopus: Access and use Support Center [WWW Document]. URL https://service.elsevier.com/app/answers/detail/a_id/14799/supporthub/scopus/ (Accessed 1.11.22).

Acknowledgment

The authors gratefully acknowledge the support of Effat College of Engineering at Effat University, Jeddah, Saudi Arabia.

References

ACTA, 2021. EHTERAZ [WWW Document]. ACTA. URL: https://www.acta.gov.qa/en/ehteraz/. (Accessed 13 May 2022).

Adadi, A., Lahmer, M., Nasiri, S., 2021. Artificial intelligence and COVID-19: a systematic umbrella review and roads ahead. J. King Saud Univ. -Comput. Inf. Sci. 34, 5898–5920.

Alcalá-Albert, G.J., Parra-González, M.E., 2021. Bibliometric analysis of scientific production on nursing research in the web of science. Educ. Sci. 11, 455.

Aljahdali, L., Sidiya, A.M., Kaki, J., Brahimi, T., 2021. Artificial intelligence and Covid-19 a bibliometric analysis. In: F the Second Asia Pacific International Conference on Industrial Engineering and Operations Management. Presented at the IEOM, Surakarta, Indonesia, pp. 2385–2393.

Almind, T., Ingwersen, P., 1997. Informetric analyses on the world wide web: methodological approaches to "webometrics". J. Doc. 53, 404–426.

Andres, A., 2009. Measuring Academic Research How to Undertake a Bibliometric Study, first ed. Chandos.

Beatty, S., 2016. Five Steps to Creating a Citation Overview in Scopus. [WWW Document]. URL: https://blog.scopus.com/posts/five-steps-to-creating-a-citation-overview-in-scopus. (Accessed 11 January 2022).

Björneborn, L., Ingwersen, P., 2001. Perspective of webometrics. Scientometrics 50, 65–82.

Brahimi, T., Bensaid, B., 2019. Smart villages and the GCC countries: policies, strategies, and implications. In: Visvizi, A., Lytras, D., M., Mudri, G. (Eds.), Smart Villages in the EU and Beyond, Emerald Studies in Politics and Technology. Emerald Publishing Limited, pp. 155–171.

Broadus, R.N., 1987. Toward a definition of "bibliometrics". Scientometrics 12, 373–379.

CAIT, 2021. Shlonik. Kuwait. Central Agency for Information Technology. [WWW Document]. App Store. URL: https://apps.apple.com/kw/app/shlonik-%D8%B4%D9%84%D9%88%D9%86%D9%83/id1503978984. (Accessed 8 May 2022).

COVID-19 Map, 2022. Johns Hopkins Coronavirus Resource Center. [WWW Document]. URL: https://coronavirus.jhu.edu/map.html. (Accessed 15 May 2022).

de Las Heras-Rosas, C., Herrera, J., Rodríguez-Fernández, M., 2021. Organisational commitment in healthcare systems: A bibliometric analysis. Int. J. Environ. Res. Public Health 18 (5), 2271.

Donthu, N., Kumar, S., Mukherjee, D., Pandey, N., Lim, W.M., 2021. How to conduct a bibliometric analysis: an overview and guidelines. J. Bus. Res. 133, 285–296.

Durou, E., Iftikhar, H., Parvez, A., Oliveira, G., 2017. National Transformation in the Middle East—A Digital Journey. Deloitte [online]. Available at: https://www2.deloitte.com/content/dam/Deloitte/xe/Documents/technology-media-telecommunications/dtme_tmt_national-transformation-in-the-middleast/National%20Transformation%20in%20the%20Middle%20East%20-%20A%20Digital%20Journey.pdf. (Accessed 11 December 2022).

Elsevier, 2022. Scopus Tutorials—Scopus: Access and use Support Center. [WWW Document]. URL: https://service.elsevier.com/app/answers/detail/a_id/14799/supporthub/scopus/. (Accessed 11 January 2022).

Falagas, M.E., Pitsouni, E.I., Malietzis, G.A., Pappas, G., 2008. Comparison of PubMed, Scopus, Web of Science, and Google Scholar: strengths and weaknesses. FASEB J. 22, 338–342.

Feuer, W., Higgins-Dunn, N., 2020. Asymptomatic Spread of Coronavirus is "Very Rare," WHO Says [WWW Document]. CNBC. URL: https://www.cnbc.com/2020/06/08/asymptomatic-coronavirus-patients-arent-spreading-new-infections-who-says.html. (Accessed 7 May 2022).

Gao, F., Jia, X., Zhao, Z., Chen, C.-C., Xu, F., Geng, Z., Song, X., 2021. Bibliometric analysis on tendency and topics of artificial intelligence over last decade. Microsyst. Technol. 27, 1545–1557.

GCC Stat, 2011. GCC Statistical Center—Home. [WWW Document]. URL: https://gccstat.org/en/. (Accessed 7 May 2022).

Goodman, J., 2022. LibGuides: Scopus LibGuide: Metrics. [WWW Document]. URL: https://elsevier.libguides.com/Scopus/metrics. (Accessed 15 May 2022).

Gunasekeran, D.V., Tseng, R.M.W.W., Tham, Y.-C., Wong, T.Y., 2021. Applications of digital health for public health responses to COVID-19: a systematic scoping review of artificial intelligence, telehealth and related technologies. NPJ Digit. Med. 4, 40.

HealthITAnalytics, 2020. White House Urges AI Experts to Develop Tools for COVID-19 Dataset [WWW Document]. HealthITAnalytics. URL: https://healthitanalytics.com/news/white-house-urges-ai-experts-to-develop-tools-for-covid-19-dataset. (Accessed 28 August 2021).

Hidayat-ur-Rehman, I., Ahmad, A., Ahmed, M., Alam, A., 2021. Mobile applications to fight against COVID-19 pandemic: the case of Saudi Arabia. TEM J. Technol. Educ. Manag. Inf., 69–77.

Holmstrom, J., 2021. From AI to Digital Transformation: The AI Readiness Framework. Business Horizons.

Hood, W.W., Wilson, C.S., 2001. The literature of bibliometrics, scientometrics, and informetrics. Scientometrics 52, 291.

Hu, Y., Yu, Z., Cheng, X., Luo, Y., Wen, C., 2020. A bibliometric analysis and visualization of medical data mining research. Medicine 99, e20338.

Huergo, J., 2020. NIST and OSTP Launch New Effort Using AI to Improve Search Engines to Aid in COVID-19 Research - ASME [WWW Document]. URL: https://www.asme.org/government-relations/capitol-update/nist-and-ostp-launch-new-effort-using-ai-to-improve-search-engines-to-aid-in-covid-19-research. (Accessed 7 May 2022).

Islam, M.M., Poly, T.N., Alsinglawi, B., Lin, L.-F., Chien, S.-C., Liu, J.-C., Jian, W.-S., 2021. Application of artificial intelligence in COVID-19 pandemic: bibliometric analysis. Healthcare 9, 441.

Ivančić, L., Milanovic Glavan, L., Vuksic, V., 2020. A Literature Review of Digital Transformation in Healthcare.

Jasarevic, T., 2020. Artificial Intelligence (AI) in Health and Six Guiding Principles for Its Design and Use. Ethics and Governance of Artificial Intelligence for Health.

Jiang, X., Coffee, M., Bari, A., Wang, J., Jiang, X., Huang, J., Shi, J., Dai, J., Cai, J., Zhang, T., Wu, Z., He, G., Huang, Y., 2020. Towards an artificial intelligence framework for data-driven prediction of coronavirus clinical severity. Comput. Mater. Contin. 62, 537–551.

Kimbrell, E., 2020. Artificial intelligence: a vital tool in the pandemic. AAAS Rep. https://www.aaas.org/news/artificial-intelligence-vital-tool-pandemic.

C. Global experiences and approaches to digital transformation in health

know.space, 2021. Bibliometric Analysis of Research Linked to UK Space Agency Funding. (CRN: 12152408; VAT: 333424820).

Krit, S., Singh, V., Elhoseny, M., Singh, Y., 2022. Artificial Intelligence Applications in a Pandemic COVID-19, first ed. CRC Press.

KSA, 2022. Saudi Vision 2030. An Ambitious Vision for an Ambitious Nation [WWW Document]. Vision 2030. URL: https://www.vision2030.gov.sa/v2030/overview/. (Accessed 15 May 2022).

Kumar, A., Gupta, P.K., Srivastava, A., 2020. A review of modern technologies for tackling COVID-19 pandemic. Diabetes Metab. Syndr. Clin. Res. Rev. 14, 569–573.

Laguarta, J., Hueto, F., Subirana, B., 2020. COVID-19 artificial intelligence diagnosis using only cough recordings. IEEE Open J. Eng. Med. Biol. 1, 275–281.

Lens, 2022. The Lens—Free & Open Patent and Scholarly Search. [WWW Document]. URL: https://www.lens.org/lens. (Accessed 10 January 2022).

MOH, 2021. Tarassud+ by Ministry of Health, Sultanate of Oman. [WWW Document]. URL: https://appadvice.com/app/tarassud/1502105746. (Accessed 10 January 2022).

MOH, 2022. BeAware Bahrain. [WWW Document]. URL: https://apps.bahrain.bh/CMSWebApplication/action/ShowAppDetailsAction?selectedAppID=321&appLanguage=en. (Accessed 10 January 2022).

Moral-Muñoz, J.A., Herrera-Viedma, E., Santisteban-Espejo, A., Cobo, M.J., 2020. Software tools for conducting bibliometric analysis in science: an up-to-date review. El profesional de la informa-ción 29 (1).

Nguyen, T.T., Nguyen, Q.V.H., Nguyen, D.T., Hsu, E.B., Yang, S., Eklund, P., 2021. Artificial intelligence in the battle against coronavirus (COVID-19): a survey and future research directions. arXiv:2008.07343 [cs].

Nobanee, H., Al Hamadi, F.Y., Abdulaziz, F.A., Abukarsh, L.S., Alqahtani, A.F., AlSubaey, S.K., Alqahtani, S.M., Almansoori, H.A., 2021. A bibliometric analysis of sustainability and risk management. Sustainability 13, 3277.

Palayew, A., Norgaard, O., Safreed-Harmon, K., Andersen, T.H., Rasmussen, L.N., Lazarus, J.V., 2020. Pandemic publishing poses a new COVID-19 challenge. Nat. Hum. Behav. 4, 666–669.

Pendlebury, D.A., 2008. Whitepaper Using Bibliometrics: A Guide to Evaluating Research Performance With Citation Data. Thomson Reuters, Place of publication not identified.

Penfold, R., 2020. Using the lens database for staff publications. J. Med. Libr. Assoc.: JMLA 108, 341–344.

Pranckutė, R., 2021. Web of science (WoS) and Scopus: the titans of bibliographic information in today's academic world. Publications 9, 12.

Pritchard, A., 1969. Statistical bibliography or bibliometrics. J. Doc. 25 (4), 348–349.

Raimondo, L., Tamimi, H., Abbas, N., 2022. 5 Keys to AI Success for GCC Health Organizations [WWW Document]. URL: https://www.boozallen.com/d/insight/blog/5-keys-to-ai-success-for-gcc-health-organizations.html. (Accessed 7 May 2022).

ResearchGate, 2020. How COVID-19 Has Affected Global Scientific Community. ResearchGate Reports. NBIC. https://statnano.com/news/67570/ResearchGate-Reports-How-COVID-19-Has-Affected-Global-Scientific-Communit.

Salmerón-Manzano, E., Manzano-Agugliaro, F., 2020. Bibliometric studies and worldwide research trends on global health. Int. J. Environ. Res. Public Health 17, 5748.

Schneegans, S., Straza, T., Lewis, J., 2022. The Race Against Time for Smarter Development, UNESCO Science Report. Licence: CC BY-SA 3.0 IGO [10330] https://unesdoc.unesco.org/ark:/48223/pf0000377433.

SDAIA, 2021. Tawakkalna. [WWW Document]. URL: https://ta.sdaia.gov.sa/en/index. (Accessed 10 January 2022).

Stephen, B., 2020. COVID-19 Databases and Journals [WWW Document]. URL: https://www.cdc.gov/library/researchguides/2019novelcoronavirus/databasesjournals.html. (Accessed 15 May 2022).

Sud, P., Thelwall, M., 2014. Evaluating altmetrics. Scientometrics 98, 1131–1143.

UAE, 2022. Al Hosn app—The Official Portal of the UAE Government [WWW Document]. URL: https://u.ae/en/information-and-services/justice-safety-and-the-law/handling-the-covid-19-outbreak/smart-solutions-to-fight-covid-19/the-alhosn-uae-app. (Accessed 13 May 2022).

Vaishya, R., Javaid, M., Khan, I.H., Haleem, A., 2020. Artificial Intelligence (AI) applications for COVID-19 pandemic. Diabetes Metab. Syndr. Clin. Res. Rev. 14, 337–339.

van Eck, N.J., Waltman, L., 2009. Vosviewer: A Computer Program for Bibliometric Mapping (SSRN Scholarly Paper No. ID 1346848). Social Science Research Network, Rochester, NY.

van Eck, N.J., Waltman, L., 2022. VOSviewer—Visualizing scientific landscapes [WWW Document]. VOSviewer. URL: https://www.vosviewer.com/. (Accessed 10 January 2022).

WHO, 2020. COVID-19 Preparedness and Response Progress. World Health Organization, Report SRP2020 https://www.who.int/publications/m/item/who-covid-19-preparedness-and-response-progress-report- - -1-february-to-30-june-2020.

WHO, 2021. COVID-19 Strategic Preparedness and Response Plan, World Health Organization, Report SRP2021. Licence: CC BY-NC-SA 3.0 IGO https://www.who.int/publications/i/item/WHO-WHE-2021.02.

WHO, 2022. The ACT-Accelerator: Two Years of Impact, World Health Organization, Report 2022. Licence: CC BY-NC-SA 3.0 IGO https://www.who.int/initiatives/act-accelerator.

Wolfram, D., 2003. Applied Informetrics for Information Retrieval Research. Greenwood Publishing Group.

Yezli, S., Khan, A., 2020. COVID-19 social distancing in the Kingdom of Saudi Arabia: Bold measures in the face of political, economic, social and religious challenges. Travel Med. Infect. Dis. 37, 101692.

Zhang, D., Saurabh, M., Brynjolfsson, E., Etchemendy, J., Ganguli, D., Grosz, B., Lyons, T., Niebles, J.C., Sellitto, M., Shoham, Y., Clark, J., Perrault, R., 2021. The AI Index 2021 Annual Report. AI Index Steering Committee, Stanford Institute for Human-Centered AI, Stanford University.

Zhang, D., Maslej, N., Brynjolfsson, E., Etchemendy, J., Lyons, T., Manyika, J., Ngo, H., Niebles, J.C., Sellitto, M., Sakhaee, E., Shoham, Y., Clark, J., Perrault, R., 2022. The AI Index 2022 Annual Report. AI Index Steering Committee, Stanford Institute for Human-Centered AI, Stanford University.

14

Digital transformation of healthcare sector in India

S.C.B. Samuel Anbu Selvan and N. Vivek

The American College, Affiliated to Madurai Kamaraj University, Madurai, Tamilnadu, India

1. Introduction

With an aim to provide better customer experience, healthcare providers recently have been trying to follow other industries and bring a fundamental shift focused toward value-based care. The rapid innovation in data, mobile, and cloud technologies has begun to disrupt the healthcare industry, forcing healthcare providers to move from a provider model to a customer-centered model catering to the needs of a modern customer who wants both choice and control. Also, organizations are ensuring that both clinical and nonclinical employees have a highly efficient work environment with information and decision control at their fingertips.

The first thing that comes to mind when we say healthcare digital transformation or a digital hospital is to go paperless. Having an integrated hospital information system (HIS) and electronic health records (EHRs) alone does not qualify as a digital hospital. For a truly digital hospital, an integrated information system has to be the base for all systems and processes.

Digital transformation is an extraordinary and accelerating transformation of business processes, activities, models, and competencies to leverage the advantages of digital technologies and their strategic impact on organizational environments, both internal and external, in a prioritized manner, keeping in mind the present and future shifts.

Every touch point that a customer or employee interacts with has got to be in sync, including searches, surgery, websites, call centers, registration, consultation, billing, admission, inpatient services, pharmacy, cafeteria, discharge, and postdischarge follow ups. The entire journey has to be taken into account while strategizing toward digital transformation.

2. ABCD approach in the digital transformation of healthcare in India

As in any industry, any change has to start from the basics, such as starting from ABCD. As a metaphor, ABCD means the foundational elements, but here it is also an acronym for the four critical elements of patient care experience. These form the most common areas of patient and employee dissatisfaction.

Appointment and admission: This means integrating all user acquisition touch points such as websites, social media, call centers, mobile apps, etc., to seamlessly facilitate appointment booking and one-time consultation or admission without the need to fill out multiple forms or redundant processes causing unnecessary delays.

Billing: Hospitals have somehow made this one single process the most complex for the patient, making the patients brand hospitals as money-minded machines. Every business needs to ask for money from its customers, but how and when to ask makes a greater difference. Integrated multipoint billing, self-pay cash kiosks, mobile pay, etc., are a few options to simplify this.

The estimates given before the procedure mostly never match the final bill and the patient is always unhappy while clearing the final bill. Hospitals have to use advanced analytics and real-time predictive data modeling to provide near-accurate estimates and thus help in better expectation management.

Clinical records: Every hospital nowadays claims to have 100% EHRs, but in reality, the quality of data they capture is pathetic; most EHRs are just scanned copies of handwritten documents. Hospitals have to realize the importance of maintaining accurate patient medical records from day 1. Along with strategic clinical and business insights, such data importantly benefit the patient and the doctor.

Discharge: Discharge is one of the processes in a hospital that, as per protocols, has to work fine but rarely does; it would be hard to find a single patient who was delighted with the discharge experience. Because discharge is a combined activity of multiple departments/processes, a delay or flaw in one of them snowballs and delays the entire process. The integrated use of HIS, EHR, inventory, billing, and insurance systems is the key. Handheld connected devices can ensure real-time updating of treatment progress notes, indenting, billing, etc. Ideally, the discharge should be done with the single click of a button.

3. Digital transformation in Indian healthcare

The world is increasingly becoming an interconnected social hub where people can quickly find means to solve complex problems through data sharing and collaboration.

Technology dominates almost every aspect of our lives. Today, we are witnessing an increasing shift toward digitalization as it makes our lives much more comfortable and convenient. Digitalization is bringing about a new revolution in almost every significant industrial sector, including business, entertainment, and science with healthcare not far behind. In fact, the healthcare industry stands on the cusp of a digital revolution that could transform the face of healthcare entirely.

Digital technologies and innovations are rapidly disrupting conventional notions of healthcare by utilizing the power of data to help doctors make better decisions, automate healthcare processes (surgery, routine check-ups, etc.) for rapid care delivery as well as promoting the overall growth of the industry.

India needs to be a part of this global hub. With the increasing growth of smartphones and rapid Internet penetration across the country, India holds vast potential to change its healthcare industry for the better. In fact, estimates suggest that India's digital connectivity will tremendously increase from 15% in 2014 to 80% by 2034, with a 58% increase in the number of rural Internet users by 2034.

3.1 Aging population and digital transformation

With the aging population steadily expanding and the rapidly increasing incidents of people suffering from chronic diseases, India urgently needs digital revolution in healthcare.

Dr. Joseph Kvedar, vice president of Connected Health, Partners Healthcare made a compelling point during his keynote speech at the Connected Health Conference in Boston he emphasized on the impact of aging population reminding us a stack reality for life-aging and implications for responsive value-based healthcare.

Globally, the aged population is on the rise because of lower birth rates and increasing life expectancy. The elderly require more attention and constant care than the younger generation. Augmented accessibility digital technologies may help the elderly population maintain good health.

For instance, fitness wearable monitors can enable the elderly to take better care of themselves. Research has established that the aged populace is likely to lead longer and healthier lives if they have a trusted social circle and feel a sense of purpose.

3.2 Lack of adequate healthcare facilities in rural areas

Even though rural areas harbor a majority of the Indian populace, most of rural India lacks the essential amenities and infrastructural facilities that are necessary for a good healthcare system.

The rural-urban divide in healthcare needs to be bridged urgently for the welfare of the large percentage of the population that resides in rural areas, people who can lose their lives due to a lack of healthcare facilities. In such a scenario, telemedicine connectivity can help the rural populace get in touch with top-tier doctors and surgeons in the country and also make treatments more cost effective.

3.3 Insufficient workforce

Surprisingly enough, India severely lags behind when it comes to meeting the bare minimum requirements of the workforce in the healthcare sector as laid down by the World Health Organization (WHO).

The inclusion of advanced digital technologies such as robotics in the healthcare system would not only make up for the workforce gap by automating various tasks such as paperwork, minor surgery, etc., but it would also decrease operational costs significantly.

Furthermore, automation in the healthcare sector would mean speedy care deliverance and increased regulatory compliance, as every step is tracked and documented.

3.4 Disease burden shift management

At present, a large part of the Indian healthcare system lacks a centralized and integrated system for storing medical records of patients. Hospitals and healthcare providers still rely on the paper-based collection of patient data, which is not only cumbersome but also time consuming.

By investing in and adopting a centralized system of retrieving population health metrics, healthcare providers can quickly identify patient groups that require immediate attention.

Furthermore, digital analytics tools such as EHRs can help store and access valuable patient data while also helping to provide faster, accurate, and holistic treatment by skipping unnecessary paperwork formalities and redundant tests.

With increasing awareness and accessibility to Internet services, today individuals are open to trying out new technologies that will enhance their life expectancy. They are more conscious, tech-savvy, and more willing than ever to embrace disruptive technologies.

Thus, the tables have turned, and the focus of healthcare has shifted from the healthcare providers to the patients.

Patients are no longer ready to be constrained by the conventional models of healthcare such as routine tests and medications. Instead, they want to take charge of their own health. Wearables are a great example. These devices can help monitor and track basic health parameters such as heart rate and blood pressure.

An example is a wristwatch that can be used as a personal emergency response system as well as a source that can relay medical data and the global positioning system (GPS) location of the patient over a remote server. This behavioral change on the part of patients is a huge driver of digitalization in the healthcare sector.

4. Digital era in Indian healthcare

The transformation of the Indian healthcare system has already begun. While the government is increasingly encouraging digitization with initiatives such as Digital India and Aadhar, the private sector has launched numerous mobile applications, telemedicine tools, and innovation centers throughout the country.

The three segments of digital technology that are helping transform the face of healthcare in India are:

4.1 M-health

Mobile health or m-health is one of the most significant parts of India's digital health revolution. Mobile healthcare apps continuously help educate individuals about preventive healthcare measures and chronic disease management. Moreover, m-health also aids in disease surveillance, tracking epidemic outbreaks, and treatment support.

In 2015, the estimated market size of m-health in India stood at 2083 crore INR, and by 2020, this number was expected to rise to 5184 crore INR. According to a study, almost 68% of doctors recommend m-health services and 59% of patients around the country are using them.

4.2 Remote diagnosis

Remote monitoring is another useful segment of digital healthcare. It allows healthcare providers to monitor patient health outside healthcare centers. It empowers patients to carry out specific routine tests using certain devices.

Wearable and mobile apps help patients get constant updates about their health. These devices help increase the rural population's access to healthcare services by offering point-of-care diagnostics, teleconsultation, and e-prescription facilities.

4.3 Telemedicine

Telemedicine uses technology for remote diagnosis, health monitoring, and consultation. It enables healthcare providers to evaluate, diagnose, and treat patients without requiring those patients to come in for an in-house visit. In addition to providing healthcare facilities in the remotest of areas, telemedicine has significantly helped bring down provider as well as patient costs (Fig. 1).

Thus, the digitalization of healthcare allows for care beyond the bounds of the hospital and into the limits of a patient's own home. With the increasing number of patients demanding healthcare services, often the infrastructure falls short. Hence, healthcare providers are now increasingly shifting toward remote monitoring services such as telemonitoring and wearables connected to the Internet of Things (IoT).

This bio-sensing wearable can track and monitor real-time changes in patient health and store the data in the health records of patients. These data can be used for early diagnosis and to provide real-time health support to patients.

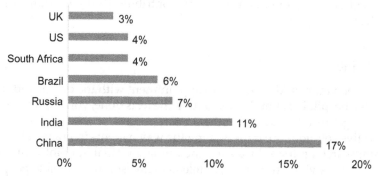

FIG. 1   Telemedicine across the world. *Source: World Bank data.*

C. Global experiences and approaches to digital transformation in health

5. Artificial intelligence (AI) in Indian healthcare

Publicly available data show that there is one doctor for every 1445 Indians, as per the country's current population estimate of 135 crore, which is lower than the WHO's prescribed norm of one doctor for 1000 people. India also ranks among the lowest with regard to hospital beds according to a WHO report, with 0.7 beds per 1000 people, far below the global average of 3.4. In addition to a shortage of healthcare professionals, compared to urban areas, rural regions face high barriers to access healthcare facilities due to poor connectivity and a limited supply of healthcare professionals.

But amid the challenges clouding the healthcare industry, there is a silver lining in the form of technology. India is among the early adopters of digital health technology. According to Philip's Future Health Index 2019 report, 76% of healthcare professionals in the country are already using digital health records. India also meets the 15 country average (46%) related to the use of AI within healthcare.

Delhi's All India Institute of Medical Science (AIIMS) hospital has a floor disinfectant and a humanoid robot in its Covid-19 wards and Fortis Hospital, Bengaluru, is using an interactive robot to screen patients and medical staff at its entrance. Measures such as telemonitored surgeries, teleeducation, telemedicine, and video consultations with doctors have been used by medical professionals during the Covid-19 pandemic.

Most of the hurdles encountered by suburbs and tier two cities today would be neutralized through the shift of the healthcare system toward collaborative and preventive healthcare. Although robotic technology is currently expensive for wider adoption across all types of healthcare settings, its adaptability specifically with reference to robotic assistants in surgery has already made inroads in India. In addition, recently many hospitals have turned to robots as maintaining social distance became the norm due to the threat posed by Covid-19.

Robots may be used to bring safer specimen processing and diagnostic procedures with zero infection risk from remote areas that normally have lower technological levels compared to modern laboratory settings.

6. Recent digital transformation in Indian healthcare

Here's how different technology trends can contribute to digital transformation in the healthcare industry:

6.1 Telemedicine

Remember when you would schedule an appointment with the doctor and wait for a couple of hours in the hospital or clinic? After getting the tests done, you would then have to wait for many days to get results and revisit the doctor.

Many innovative solutions are transforming the way patients interact with healthcare professionals. From searching for a doctor to scheduling a virtual appointment and communicating with doctors via video or voice call, telemedicine solutions enable people's access to health professionals on demand.

Telemedicine evolution is one of the most significant transformations in the US healthcare market. In a big country such as the United States, where access to healthcare providers is limited, telemedicine is emerging vigorously. As much as 90% of surveyed healthcare executives revealed that organizations started building or integrating a telemedicine system.

One of the best examples of telehealth technology is virtual appointments between patients and doctors. In remote or rural areas where access to healthcare is limited, virtual appointments facilitate patients communicating with doctors. Patients facing mobility challenges can use telemedicine to interact with health professionals.

Telehealth technology can also be used to manage patients at high risk and enable health professionals to track the patient's conditions and activities remotely via IoT-based health sensors and wearable devices.

It is essential to consider that any telemedicine app or solution should comply with legislation in your targeted regions or countries. We have built a telemedicine app for healthcare institutes that facilitates doctors interacting with both existing and new patients and patients communicating with existing doctors via video, voice, or text chat.

6.2 Using big data in healthcare

Big data is transforming the way we analyze, leverage, and manage data in every industry. Healthcare is one of the promising industries where it can be implemented to avoid preventable diseases, enhance the quality of life, reduce treatment costs, and forecast outbreaks of epidemics.

Health professionals can collect a massive amount of data and find the best strategies to use the data. Using big data in healthcare can have positive and life-saving outcomes.

With emerging technologies, it has become easier to not only collect essential healthcare data but also convert it into valuable insights to provide better care. Using data-driven insights, health professionals can predict and solve an issue before it becomes too late.

Let's understand how big data can be used in healthcare and what benefits it provides.

- *Patient prediction for improved staffing*

    A healthcare shift manager usually faces an issue of how many people should be on staff at any specific time. If a manager keeps too many workers, there may be the risk of unwanted labor costs and resources. On the other side, having too few workers can also result in poor customer service outcomes that can be riskier for a patient's health.

    Big data can solve this issue. Data from a wide array of sources can be used to generate daily and hourly predictions of how many patients are expected to be at the hospital or clinic. In Paris, four hospitals that are part of the Assistance Publique-Hôpitaux de Paris have used a variety of sources, including 10 years of hospital admission records, to provide daily and hourly predictions of how many patients are expected to be at the hospital at any specific time.

    Therefore, collecting data and using it to discover patterns to predict behavior can help improve staffing by predicting patient admission rates.
- *Real-time alerting*

    Real-time alerting is also one of the crucial examples of big data analytics in the healthcare industry. Hospitals use clinical decision support software to analyze medical data on the spot and provide health practitioners with advice to help them make informed decisions.

Wearable devices are used to collect patient health data continuously and send it to the cloud. For example, if a patient's heartbeat increases suddenly, the system sends a real-time notification to the doctor who can then take action to lower the rate and reach the patient.

Because IoT devices generate a massive amount of data, implementing intelligence on the data can help health professionals make important decisions and get real-time alerts.

- *Informed strategic planning using health data*

  Big data in healthcare facilitates strategic planning. Healthcare managers can analyze the results of patient check-ups in various demographics groups. Also, they can find factors that discourage people from taking up treatment.

  The University of Florida used free global health data and Google Maps to create heat maps targeted at specific issues such as chronic diseases and population growth.

  Therefore, healthcare data can also be used for planning informed strategies.

- *Preventing human errors*

  Many times, it has been found that the professionals either send a different medication or prescribe the wrong medicine by mistake. Big data can be used to reduce such errors by analyzing prescribed medicine and user data.

  Prescription data collected from different medical professionals can be monitored using the big data healthcare tool. The software can flag prescription mistakes made by any physicians and help save many lives.

6.3 Internet of Things (IoT)

Before the introduction of the IoT, patient and doctor interactions were restricted to physical visits and text communications. Doctors or hospitals had no way to track the patient's health continuously and take action accordingly.

IoT-enabled devices facilitate remote monitoring in the healthcare industry, unlocking the potential to keep patients healthy and safe and allowing physicians to provide better care. Because IoT has made interactions with doctors efficient and easier, it has improved patient satisfaction and engagement.

Also, remote monitoring of patient health helps in preventing readmissions and decreasing the duration of stay in the hospital. IoT can also reduce healthcare costs and improve treatment outcomes.

IoT is changing the healthcare industry by remodeling people's interactions in providing healthcare solutions. The implementation of IoT in healthcare benefits physicians, hospitals, patients, and insurance companies.

- *IoT for physicians*

  With home monitoring equipment embedded with IoT sensors as well as wearable devices, physicians can monitor patient health in real time. IoT allows healthcare professionals to become more watchful and interact with patients proactively. Data gathered from IoT devices can help doctors find the best treatment process for patients and get the expected outcomes.

- *IoT for patients*

    Devices such as fitness bands and wirelessly connected heart rate monitoring cuffs provide patients access to personalized attention. IoT devices are used to remind of doctor appointments, count calories and the number of steps taken in a day, monitor blood pressure and heart rate, and much more.

    IoT enables real-time remote monitoring and is beneficial for elderly patients. It uses an alert mechanism and sends a notification to concerned healthcare providers and family members.

- *IoT for hospitals and clinics*

    Apart from tracking patient health, IoT devices can be used in many other areas in hospitals. IoT devices embedded with sensors are used for monitoring the real-time location of medical equipment, including nebulizers, wheelchairs, oxygen pumps, and other equipment.

    Hospitals also have to deal with the spread of infection, which is a primary concern. IoT-based hygiene monitoring devices assist in preventing patients from getting infections. For example, smart IoT-enabled cameras can detect whether patients are washing or sanitizing their hands before taking a meal or medication or visitors are sitting too close to the patient.

    Also, IoT devices can help in managing assets, for example, monitoring refrigerator temperature and humidity.

- *IoT for health insurance companies*

    With IoT-connected devices, health insurers can seamlessly perform their underwriting and claims operations. Data captured from health monitoring devices can be used by insurance companies to identify fraudulent claims and detect underwriting prospects.

    IoT devices introduce transparency between customers and insurers in the underwriting, claims management, risk assessment process, and pricing.

    Insurers can retain customers by rewarding them for using and sharing health data captured by IoT devices. They can give incentives if a person keeps track of their routine activities and maintains a healthy lifestyle. It will help insurers reduce claims drastically.

    Insurers can also verify claims with the help of data generated from IoT devices.

IoT can contribute to digital transformation in healthcare with the massive amount of data generated by IoT devices. IoT technology has a simple four-step architecture that can apply to any of the industries.

6.4 Virtual reality (VR)

Virtual reality is a technology that uses the computer-generated simulation of a three-dimensional image or environment to allow a person to hear, see, and interact using special equipment such as headsets.

The technology creates a simulated environment where users can be immersed. Unlike traditional user interfaces, VR takes users inside a virtual experience instead of only displaying a screen.

The healthcare industry is adopting virtual reality to deliver better care to patients. For example, one of the patients was getting chemotherapy every week for around 6 years to treat colon cancer. She used to spend her 4.5 h during the chemo session reading books, chatting, or watching TV.

During infusion, she sometimes wanted to go to beaches to relax. Unfortunately, she was unable to go in real life as her skin was too sensitive to go out in the sunlight. But VR made her dream come true by simulating a beach-like environment where she could feel like she was sitting on the beach and enjoying the sun.

She is not the only one who is fond of using virtual reality in a healthcare setting, as many patients love this experience when getting treated.

From the clinic to medical rooms, VR is exploding and expected to continue to grow in the coming years. According to research by GlobeNewsWire, the market for VR in healthcare will reach $7 billion by 2026.

Healthcare is still in its early stages of the technology; therefore, the healthcare industry has started to realize where it can be used and the challenges posed by VR.

VR can help the healthcare industry in the following ways:

- *Pain reduction*

  Children's National Hospital in Washington, D.C., piloted a program in 2019 that included a virtual reality headset for children who required procedures such as removing stitches, sutures, and foreign bodies in emergency rooms.

  The program involved around 40 children aged 7–23. Each child was provided a VR headset covered in protective gear to cut down on germs. Then, children selected different scenarios, including walking through a jungle, talking with a friendly snake, and taking a roller coaster ride. VR headsets were connected to screens so that parents could see what their children were watching.

  During the whole activity, kids did not experience any pain and parents were thrilled as their child was happy and tolerated the procedure peacefully.

  Therefore, many healthcare institutes are adopting VR technology for pain reduction therapies.
- *Pacing recovery in physical therapy*

  With VR, you can make physical therapy for patients more enjoyable by engaging them in a simulated environment. One of the studies revealed that children with cerebral palsy saw a significant improvement in their mobility after VR therapy.

  Neuro Rehab VR, one of the leading providers in VR physical therapy, has introduced a gamified approach to physical therapy. The company has developed VR training exercises with machine learning to customize each exercise to the patient's therapeutic requirements.

  That's how VR is being used widely in speeding recovery in physical therapy.

6.5 Artificial intelligence (AI)

Artificial Intelligence simplifies the lives of doctors, patients, and hospital administrators by doing tasks that are usually done by humans at a fraction of cost and in less time.

From finding links between genetic codes to driving surgery-assisting robots, surveying chronic diseases and conducting risk assessments, AI is reinventing and revitalizing modern healthcare through machines that can comprehend, predict, learn, and act.

AI provides a number of advantages over clinical decision making and traditional analytics. Learning algorithms can become more accurate and precise when they interact with training data. AI allows humans to gain unprecedented insights into care processes, treatment variability, patient outcomes, and diagnostics.

AI is poised to bring digital transformation in healthcare in the following ways:

- *Diagnosing and reducing errors*

    Medical errors and misdiagnosing illnesses resulted in 10% of all deaths in India in the last years. AI is one of the most exciting technologies that promises to improve diagnostic processes.

    Large caseloads and incomplete medical histories can result in deadly human errors. However, AI can help predict and diagnose diseases faster than medical professionals. For example, in one study, an AI model used algorithms and deep learning to diagnose breast cancer at a higher rate than 11 pathologists.

    Breast cancer is found to be the second-leading cause of cancer deaths among women in the United States, and mammographies have reduced mortality.

    Computer-assisted detection and diagnosis (CAD) software was built in the 1990s to help radiologists enhance the predictive analytics of mammographies. Unluckily, data suggested that early CAD systems did not led to an improvement in performance.

    However, the remarkable success of deep learning in visual object detection and recognition as well as deep learning tools assisted radiologists in improving the accuracy of mammographies.

- *Analytics for pathology images*

    Pathologists provide one of the essential sources of diagnostic data for providers across the spectrum of care delivery. According to Dr. Jeffrey Golden, Chair of the Department of Pathology and a Professor of Pathology at Harvard Medical School, "Seventy percent of all decisions in healthcare are based on a pathology result. Somewhere between 70% and 75% of all the data in an electronic health record (EHR) are from a pathology result. So the more accurate we get, and the sooner we get to the right diagnosis, the better we're going to be. That's what digital pathology and AI (have) the opportunity to deliver."

    One of the digital pathology platforms, Proscia, uses AI to identify patterns in cancer cells. It helps pathologists remove bottlenecks from data management and leverage AI-enabled image analysis to link data points that support cancer diagnosis and treatment.

    AI can also enhance productivity by exploring features of interest in slides before a clinician reviews the data.

- *Converting smartphone selfies into powerful diagnostic tools*

    Harnessing the power of portable devices, experts assume that images captured from smartphones and other sources can be an essential supplement to medical quality imaging, especially in developing nations. Because the quality of smartphones is improving every year, phones can produce images that are viable for analysis by AI algorithms. Some researchers in the United Kingdom have built a tool that recognizes developmental diseases by analyzing a child's face image. The algorithm can identify discrete features, such as the child's eye and nose placement, jaw line, and other attributes that indicate a craniofacial abnormality. Therefore, it's an excellent opportunity for us to convert a lot of data into valuable insights.

    Smartphones can be used to collect images of skin lesions, wounds, medications, infections, or eyes to help underserved areas manage a shortage of specialists while reducing the time to diagnosis for some complaints.

C. Global experiences and approaches to digital transformation in health

- *Management of electronic health records*

    Data (including patient information, new research findings, and diagnosis details), also called EHRs, are generated in large volumes every day in the healthcare industry. The implementation of AI in EHRs helps organizations gain insights to collaborate with patients and make informed decisions. EHR systems, combined with AI, can provide healthcare professionals with the ability to manage their observations instead of adding the data manually. In EHRs, AI provides healthcare professionals the opportunity to examine the existing data and extract significant insights they can use to give recommendations.

    AI facilitates healthcare professionals to leverage the information in EHRs, transforming them into virtual assistants that can deliver value to healthcare professionals.

    With AI-powered EHR, healthcare professionals can get notifications about things they should consider when recommending a drug or treatment. For example, if a medicine is not good for a patient based on the profiling of the patient's genes, the AI-based system can provide a more relevant recommendation.

- *Drug and vaccine creation*

    Before understanding how AI contributes to the development of drugs, let's understand the cycle of drug development. Researchers find a target protein that's causing the disease. They examine such proteins diligently for a long time. Otherwise, there can be a significant risk of losing a lot of money on the wrong protein. Then, the research team tries to explore a molecule or a compound that would affect the protein. The compound should be able to modify the protein to determine the disease-causing protein effectively.

    During the process, ineffective compounds are thrown aside and only safe and efficient compounds are taken for drug development. The entire process is manual and time consuming; therefore, AI comes into the picture.

    As there are hundreds and thousands of molecules out there, human researchers are not able to test each of these molecules manually. Yet, without testing each of the molecules, it is not possible to determine which molecule would be the most relevant to fight a specific disease.

    Using AI, experts would feed in the parameters. They search through all the molecules and every molecule is compared against parameters. The AI-enabled system will keep learning from the generated data and find one or more compounds that are most equipped to fight the disease.

    Similarly, vaccines can be developed and tested successfully with the help of artificial intelligence.

- *Automating repetitive processes*

    AI technology is poised to automate repetitive tasks of the healthcare industry, setting administrators free to work on higher-level ones. From eligibility checks to data migrations and nonjudicial claims, everything can be automated so that staffers can work on offering better patient service. Olive, one of the AI-as-a-service tools, can be integrated easily into a hospital's existing software, removing the need for expensive downtimes or integrations.

- *Notifying doctors when patients are in trouble*

    Many hospitals all over the world are using Google's Deepmind Health AI to help drive patients from diagnosing to treatment efficiently. The Deepmind program alerts doctors when a patient's health worsens and can even assist in the diagnosis of diseases by searching through the massive dataset for related symptoms. By gathering patient symptoms and entering them into the Deepmind platform, doctors can detect a disease more efficiently and quickly.

6.6 Cloud access and mobility

Data are of less importance if they cannot be accessed in real time. With digitization, the quantity of data captured is enormous, and the dependency on this data is high too. Mobility and cloud access makes data storage and access very efficient. Doctors and patients are becoming more and more comfortable in accessing healthcare information on mobile devices. Paper charts and document rooms will soon be history. Patients, hospitals, and insurance companies are now storing medical records in the cloud, with the ability to access data online 24/7.

6.7 Alexa

The voice assistant from Amazon has already been successful in the lifestyles of people and smart offices. In the next few years, nurses, doctors, patients, and pharmacists will be using Alexa in their professions.

Voice technology like Alexa can enable communication between healthcare providers and patients.

- *Enabling diabetes patients to manage their solution effectively*

  AWS and Merck together initiated the Alexa Diabetes Challenge in 2018 with a $250,000 prize. Wellpepper became the winner with its Sugarpod, a clinically tested and Alexa-based digital platform for the management of diabetes.

  Using Sugarpod, Alexa can help diabetics handle their treatment and track progress effectively. It shows how Alexa can be implemented to transform chronic diseases such as type 2 diabetes.
- *Enhancing interaction in the hospital*

  One of the start-ups in Los Angeles, Aiva, uses an Alexa-enabled platform to help patients communicate with caretakers such as nurses and manage in-house entertainment. With this platform, patients can ask Alexa to switch on/off the television, change the channel, and call the caretakers using a mobile phone. It facilitates patients seeking quick assistance, and many experts think this interactive technology can also eliminate loneliness for patients.
- *Managing blood pressure with Alexa Skills*

  At a time when life seems stressful, one of the popular medical equipment companies, Omron, has launched a watch called HeartGuide that can measure blood pressure and send readings with Alexa Skills.
- *Reducing wait times*

  Hospitals can enable Alexa Skills using which patients can fetch contact information for particular departments. Using that Alexa Skill can save both time and stress by informing patients about the wait time in each clinic or hospital. It helps the patient be on time and prevents critical situations.

6.8 Robots on the surgical beat

India's first robotic-assisted surgical procedure took place at a Delhi hospital in 2002. The world's first heart surgery from a remote location was done in India in 2018 by Gujarat's Dr. Tejas Patel, a cardiac surgeon. With the help of advanced robotics, Patel conducted the world's first telerobotic surgery on a middle-aged women with a blocked artery while sitting 32 km away from the patient.

Over the years, robotic-assisted surgeries have made significant contributions to Indian healthcare, with the growth of the Indian robotic-assisted surgical market expected to reach Rs.2600 crore by 2024 at a CAGR of 19.8%, as per a report by Research and Markets. Based on the estimates from a symposium in 2021, almost 50% of all surgeries in India will be robot assisted by 2025. Concurrently, research and development in the fields of surgical systems, healthcare products, and diagnostics will also see a marked uptick.

There are reports that more than 500 robotic surgeons exist in India, including deployment at government and private healthcare facilities. Due to a shorter recovery period after surgery as well as relatively less pain and blood loss, robotic-assisted surgeries are considered to be a better alternative to open surgeries and laparoscopic surgeries.

Wider use of robotic technology will enable Indians to save traveling and boarding costs over and above the usual hospitalization costs. Specifically, in the case of medical surgery, robotics have been facilitating experts, the majority of whom are concentrated in metropolitan areas and not available due to remote and inaccessible terrain for the tier two and three populations.

Recently, the world's first cost-effective robotic endo training kits were developed in India in collaboration with Korea. It is a surgical training robot equipped to provide a high-definition observation of the patient's anatomy. The live feed of the anatomy is broadcast on a screen, which leverages virtual reality technology.

In addition to assisting surgeons, robots have been playing a critical role in training young medical professionals to become future surgeons. A cost-effective robotic surgical training has become an ideal training model for young surgeons in tier two and three hospitals. It has accelerated the process of making young surgeons more skilled by reducing time and enhancing effectiveness.

Given the vast gap in the doctor-patient ratio, advanced technologies such as AI and robotics are expected to play a key role in facilitating important medical services in the country, especially in the middle of a pandemic. They not only lend support in hospitals and nursing homes, but also in the development, testing, and production of medicine, vaccines, and other medical devices and auxiliaries. Here are some fast-growing developments that will further shape the healthcare robotics industry in the years to come.

Nano robots set to dominate: With a microscopic virus holding humanity to ransom, the near future will surely see a proliferation of nanoscale robots who will traverse through the body to detect anomalies and irregularities. While the technology around nanoparticles is still developing, nanotechnology is a field that has the potential to revolutionize the future of healthcare and medicines.

Popularity of rehabilitation robots: With longevity on the rise and an expanded base of senior citizens in almost all large economies, the development of innovative rehabilitation robots is set to boom. Such robots will support an aging population grappling with physical disabilities and patients through the use of integrated sensors.

Surgical robotic systems: This is expected to be particularly relevant to hospitals that offer surgical procedures for complicated conditions. Moreover, with enhancement of surgical systems being a pertinent need, it will drive the market for robotic services. Low-cost innovations that assist robotic surgeries are expected to dominate the Indian market.

Training surgical specialists to become proficient in clinical robotics is a welcome trend that is sharply on the rise. Advances in surgical platforms and innovations propelled by VR/AR have the potential to address the needs of the fast-growing clinical robotic sector and usher in a new era of robot-assisted medical care for the masses. The journey of robotics from industrial assembly lines to suburban and small town clinics promises to be an exciting one.

7. Case study: Apollo hospitals launch robot-assisted cardiac surgery unit

The cardiothoracic surgeon Dr. M.M. Yusuf and his team at Apollo Hospitals, Chennai, continue to perform this latest procedure successfully with better outcomes. The risk involved when compared to conventional surgery is less and it gives a new lease on life to the patient.

This surgery is performed under general anesthesia. The cardiac surgeon will make tiny holes in the left chest cavity. The surgical instruments are attached to robotic arms and a camera is inserted through these holes. The cardiac surgeon will control the robotic arms and camera from a console located in the same operation theater. Once the surgery is complete, the surgeon will remove all instruments and close the incisions.

This unique procedure is performed in very few centers worldwide. At Apollo Hospitals, Yusuf and his team have been performing this procedure successfully.

After surgery, patients will be monitored for recovery. Patients generally have less blood loss, minimal pain, and faster recovery. Based on the recovery, patients will be discharged 2–3 days after surgery (The Hindu dated July 28, 2021).

8. Way forward in digital health sector

The National Digital Health Mission (NDHM) envisions to "create a national digital health ecosystem that supports universal health coverage," but there are operational areas that need capacity building without which the exercise might prove costly in terms of acceptance, usability, and data aggregation. The capacity building activities should be implemented considering the intrinsic and contextual factors. The operational areas and opportunities for capacity building are highlighted in Table 1.

A recent report suggests that strong governance is the key to digital health investments and advocates for the use of governance frameworks such as the Health ICT Governance

TABLE 1   National digital health mission components and opportunities for capacity building.

NDHM components	Capacity building areas
Health ID	Promoting digital health literacy among patients and providers to enhance communication channels. This promotion should focus on overall digital literacy and should be based on a competency-based framework (Haskew et al., 2015)
PHR	Promoting private healthcare enterprises to actively engage in early adoption of the same as evidence suggests that private providers have often taken a back seat in reporting of health management information system and integrated disease surveillance program (Health IT Analytics, n.d.)
EHR web applications	Evidence suggests that EHRs have not been well accepted among healthcare professionals globally owing to high patient volumes and poor change management mechanism (Health Nucleus, n.d.). This warrants a strategy to foster a smooth transition from paper-based medical records to web-enabled applications at a country level, taking diversity and infrastructural limitations into consideration
Digital doctor platform (doctor's directory)	Indigenous and traditional medicines require careful integration into the national digital health architecture and in doing so, India will be truly able to demonstrate interoperability
Health Facility Registry	The Health Facility Registry should provide information to patients about all the possible diagnostic tests and services available for different schemes such as CGHS, Ayushman Bharat, and other government-initiated schemes because empaneled hospitals are dissatisfied with low tariffs and sometimes do not offer the required service to the patients. This would ensure transparency among the stakeholders (patients, providers, and the government)

CGHS, *central government health services*; EHR, *electronic health record*; NDHM, *National Digital Health Mission*; PHR, *personal health record*.
Source: *ndhm.gov.in*.

Architecture Framework 2.0 and Control Objectives for Information and Related Technologies 5 (COBIT-5). Lessons from other countries have identified cross-cutting challenges such as sustainable funding, training of users, and establishing credibility in information handling. Engaging all the actors (boundary spanners and gatekeepers) in the digital health sphere with due attention given on "what to govern" and how to deploy digital governance (top-down and bottom-up) across various levels of health care would bring us a step closer to the holistic digital health ecosystem in India.

The sandbox environment at the NDHM will allow one to integrate and validate software systems by partnering with the NDHM (application programming interfaces [APIs]). Enrollment is open to healthcare service providers, hospitals, healthcare software vendors, and anyone who wants to build NDHM APIs. Health information providers, health repository providers, health information users, or health lockers should integrate with NDHM APIs. Table 2 shows the digital building blocks hosted by the sandbox (ndhm.gov.in).

TABLE 2 Digital building blocks hosted by the sandbox.

Digital building blocks	NDHM services
Health ID service and APIs	For creating a sandbox Health ID and integrating the software with Health ID APIs
Consent manager and gateway	For registering software to test link records and process health data consent requests
PHR mobile applications (android)	To manage one's Health ID, health records, and consents
Health information user applications	To create consent requests for Health ID

APIs, *application programming interfaces*; NDHM, *National Digital Health Mission*; PHR, *personal health record*.
Source: *ndhm.gov.in.*

9. Challenges in the digital transformation of the health sector in India

As with any change management, digital transformation has its own set of challenges. Experts would categorize these into three areas: Technology, operations, and cost.

Technology: Integrating with multiple IT systems, finding a qualified implementation partner, legacy issues, and very few single platform solutions in the market.

Operations: Cultural change, especially with senior doctors, ground staff, and as an irony the IT department who have to drive this initiative. All will have a tough time in subscribing to this concept.

Cost: Digital transformation is cost intensive and has to be considered as an expense toward digital infrastructure to develop, maintain, and upgrade.

10. Digital milestones in Indian healthcare

10.1 2015 Digital India campaign

The Digital India program was launched by the Indian government with the goal of altering the country's economy. This includes making sure that all government services, such as healthcare, are accessible to residents via electronic means. To facilitate this, upgrades to Wi-Fi availability and the introduction of 5G are planned.

10.2 2017 Digital health identity document (IDs)

A digital health ID is proposed as part of a digital healthcare system that integrates data for citizens and stakeholders from private and public healthcare providers. Hospitals, Internet pharmacies, telemedicine companies, laboratories, and insurance companies are expected to participate in the new system.

10.3 2020 COVID-19 and the acceleration of digital healthcare

Between March and May 2020, Covid-19 led to a 500% rise in the use of virtual consultations. The pandemic, however, revealed a digital divide: in order to book vaccines, people must go to the CoWin website and upload their identity documents using a smartphone or computer with a good Internet connection, which is available in urban areas but not in India's vast rural areas.

10.4 2021 National Digital Health Mission officially launched

In September, India's National Digital Health Mission was launched. Other projects include a digital registry of doctors dubbed Digi-doctor and EHRs, in addition to the larger implementation of IDs. A planned unified health interface also aspires to be an open, interoperable platform that connects various digital health technologies.

11. Summary

The digital transformation of India's healthcare business is not just beginning; it has already begun. However, putting a method to this chaos will necessitate a significant amount of strategizing and planning. They will have to embrace open technologies that enable advanced data analysis and the development of customer-centric services. They must be able to see processes as continuous flows rather than discrete handoffs, accept greater risk, move faster, and form new alliances.

The achievers in Indian healthcare are implementing change swiftly and capitalizing on the previously mentioned method. They're making early investments in potential technology and risk-sharing partnerships with companies both inside and outside the sector. They're adopting new development and operational methods, and they're relying increasingly on data-driven insights to make key business choices. More crucially, they are reimagining themselves as digital enterprises—adaptive, collaborative organizations that can keep up with market developments. This enables them to successfully complete their transformation programs, resulting in improved patient outcomes and increased value for all stakeholders.

12. Conclusion

Technology will address the challenges of traditional systems by streamlining the processes. The healthcare space sets an example of a sector enacting massive changes. Digitization has penetrated our daily lives and the time is ripe to leverage technology that will create patient-centric healthcare systems that can improve response time, reduce human error, and save costs. Technology is bridging the gap and altering the way doctors and patients interact with each other. The transformation healthcare services today are witnessing a revolution with advanced medical technology as well as improved and precision diagnostic and therapeutic tools.

The industry is getting a lot more organized and becoming more patient-centric. According to the National Association of Software and Service Companies findings, the Indian healthcare IT market was about $1 billion in 2014 and was pegged to grow around 1.5 times by 2020. Currently, the healthcare software market comprises only 9% of the total healthcare

IT market in the country. This is not all; around 60% of the health-tech focused start-ups have incorporated over the last 8 years.

By adopting digital technology and processes, healthcare providers can focus on improving the quality of patient care, but this is not restricted to just adopting a new technology. There is a need to have a proper digital strategy and required IT infrastructure. Here are some ways technology could help us live healthier and more productive lives.

electronic health records (EHRs) helps track patient health data and support medical decisions. Cloud computing is becoming popular for facilitating EHRs sharing and integration. But adopting such technology could be nerve-wracking, as it entails greater responsibility. Information security becomes an important issue when moving EHRs to the cloud environment. Telemedicine is another exciting area with well-designed solutions to deliver timely and effective healthcare. Consultation through mobile devices using video, images, and audio channels can help patients access good medical aid. A study by ASSOCHAM highlights India's telemedicine market potential to cross $32 million by 2020 at a CAGR of more than 20%.

References

Haskew, J., Rø, G., Saito, K., Turner, K., Odhiambo, G., Wamae, A., Sharif, S. and Sugishita, T., 2015. 'Implementation of a cloud-based electronic medical record for maternal and child health in rural Kenya'. Int. J. Med. Inform., 84(5), pp. 349–54. https://doi.org/10.1016/j.ijmedinf.2015.01.005. (Accessed 30 March 2020).

Health IT Analytics (n.d.). Google Pursues AI, Voice Recognition for EHR Workflows'. https://healthitanalytics.com/news/google-pursues-ai-voice-recognition-for-ehr-workflows. (Accessed 5 March 2020).

Health Nucleus (n.d.). 'Artificial Intelligence & Machine Learning'.

Recommended readings

https://www.dqindia.com/digital-transformation-can-accelerate-growth-indian-healthcare-landscape-avenues/.

https://smefutures.com/digital-health-the-new-rx-for-indian-healthcare-ecosystem/.

https://www.timesnownews.com/india/video/mapping-the-digital-transformation-in-the-healthcare-sector-on-digital-india-summit/694951.

https://www.ey.com/en_gl/life-sciences/how-digital-transformation-can-benefit-the-entire-health-care-ecosystem.

Recommended videos

https://www.youtube.com/watch?v=ip2Lp12SoL8.
https://www.youtube.com/watch?v=KiiXrnpGvl4.
https://www.youtube.com/watch?v=Xw_71wud7og.
https://www.youtube.com/watch?v=cM4aep7VXb8.
https://www.youtube.com/watch?v=wO_1GxWihzw.
https://www.youtube.com/watch?v=jJwdT55sEDU.
https://www.youtube.com/watch?v=aQ8HEqt7CCs.

Further reading

Accenture, 2017. Artificial Intelligence: Healthcare's New Nervous System. https://www.accenture.com/_acnmedia/PDF-49/Accenture-Health-Artificial-Intelligence.pdf#zoom=50. (Accessed 5 March 2020).

Ash, M., Petro, J., Rab, S., 2019. How AI in the exam room could reduce physician burnout. Harv. Bus. Rev. 12 November 2019. https://hbr.org/2019/11/how-ai-in-the-exam-room-could-reduce-physician-burnout. (Accessed 5 March 2020).

C. Global experiences and approaches to digital transformation in health

Bailey, R., Parsheera, S., Rahman, F., Sane, R., 2018. Disclosures in privacy policies: does "notice and consent" work? In: Working Paper, No. 246. National Institute of Public Finance and Policy, New Delhi. https://www.nipfp.org.in/media/medialibrary/2018/12/WP_246.pdf. (Accessed 10 March 2020).

Business Line, 2018. 1 bn Records Compromised in Aadhaar Breach Since January: Gemalto. The Hindu Business Line. 20 October 2018. https://www.thehindubusinessline.com/news/1-bn-records-compromised-in-aadhaar-breach-since-january-gemalto/article25224758.ece. (Accessed 11 March 2020).

Central Bureau of Health Intelligence, 2018. National Health Profile 2018. Ministry of Health and Family Welfare. http://www.cbhidghs.nic.in/WriteReadData/l892s/Before%20Chapter1.pdf. (Accessed 10 March 2020).

Colclough, G., Dorling, G., Riahi, F., Ghafur, S., Sheikh, A., 2018. Harnessing Data Science and AI in Healthcare: From Policy to Practice, Report of the WISH Data Science and AI Forum 2018. https://www.wish.org.qa/wp-content/uploads/2018/11/IMPJ6078-WISH-2018-Data-Science-181015.pdf. (Accessed 30 March 2020).

Culnane, C., Rubinstein, B.I.P. and Teague, V. (2017), 'Health Data in an Open World, arXiv preprint, 15 December 2017, arXiv:1712.05627 (Accessed 10 March 2020).

Dargan, R., 2019. AI Applications Aid Efficiency, Ease Workload. Radiological Society of North America. 16 July 2019. https://www.rsna.org/news/2019/july/ai-applications-address-workload. (Accessed 5 March 2020).

Das, M., Angeli, F., Krumeich, A.J.S.M., van Schayck, O.C.P., 2018. The gendered experience with respect to health-seeking behaviour in an urban slum of Kolkata, India. Int. J. Equity Health 17 (1), 24. https://www.ncbi.nlm.nih.gov/pmc/articles/PMC5813424/#__ffn_sectitle. (Accessed 11 March 2020).

Davenport, T. and Kalakota, R. (2019), 'The potential for artificial intelligence in healthcare', Future Healthc. J., 6(2): pp. 94–98, https://doi.org/10.7861/futurehosp.6-2-94 (Accessed 19 May 2020).

Deo, S., Tyagi, H., 2019. Rx ICT: Digitally Disrupting Healthcare. Forbes India. 22 February 2019. http://www.forbesindia.com/article/isbinsight/rx-ict-digitally-disrupting-healthcare/52629/1. (Accessed 10 March 2020).

ET Rise, 2018. How Manthana uses AI to process visual medical data. Economic Times. 17 August 2018. https://economictimes.indiatimes.com/small-biz/startups/newsbuzz/how-manthana-uses-ai-to-process-visual-medical-data/articleshow/65443587.cms. (Accessed 11 March 2020).

Expert System, 2017. What Is Machine Learning? A Definition. 7 March 2017. https://www.expertsystem.com/machine-learning-definition. (Accessed 5 March 2020).

Express News Service, 2016. Maharashtra website hacked: diagnostic lab details of 35,000 patients leaked. The Indian Express. 3 December 2016. https://indianexpress.com/article/india/diagnostic-lab-details-of-35000-patients-leaked-hiv-reports-4407762. (Accessed 10 March 2020).

Forbes, 2019. AI and Healthcare—A Giant Opportunity. 11 February 2019. https://www.forbes.com/sites/insights-intelai/2019/02/11/ai-and-healthcare-a-giant-opportunity/#60febfc4c682. (Accessed 5 March 2020).

Garrett, P., Seidman, J., 2011. EMR vs EHR—What is the Difference? Health IT Buzz. 4 January 2011. https://www.healthit.gov/buzz-blog/electronic-health-and-medical-records/emr-vs-ehr-difference. (Accessed 19 May 2020).

Gershgorn, D., 2018. If AI is going to be the world's doctor, it needs better textbooks. Quartz. 6 September 2018. https://qz.com/1367177/if-ai-is-going-to-be-the-worlds-doctor-it-needs-better-textbooks. (Accessed 11 March 2020).

Goyal, M., 2019. How Wadhwani brothers Sunil and Romesh are using AI to serve the underserved. The Economic Times. 5 March 2019. https://economictimes.indiatimes.com/tech/software/how-wadhwani-brothers-sunil-and-romesh-are-using-ai-to-serve-the-underserved/articleshow/68268245.cms. (Accessed 11 March 2020).

Gupta, P., 2018. Bringing Together an AI Network for Eyecare to Prevent Avoidable Visual Impairment. The Microsoft India Blog. 14 March 2018. https://docs.microsoft.com/en-us/archive/blogs/msind/mine-ai-network-for-eyecare. (Accessed 11 March 2020).

Hart, D.R., 2017. If you're not a white male, artificial intelligence's use in healthcare could be dangerous. Quartz. 10 July 2017. https://qz.com/1023448/if-youre-not-a-white-male-artificial-intelligences-use-in-healthcare-could-be-dangerous. (Accessed 11 March 2020).

Investment in healthcare and medical technology in Kuwait

Ahmad Salman[a], Ali Al-Hemoud[b], Hasan Al-Attar[b],
and Mariam Malek[b]

[a]Kuwait Ministry of Health, Safat, Kuwait [b]Environment and Life Sciences Research Center,
Kuwait Institute for Scientific Research, Safat, Kuwait

1. Introduction

Although it is established that healthcare is the most promising field for economic growth and development in Kuwait, the country primarily relies on external resources to meet the needs of the medical technology and healthcare industries. The health sector in Kuwait lacks knowledge and expertise. Owing to a lack of investment in medical technology, there is growing reliance on foreign resources and labor, thereby leading to increasing distrust in local treatment and medical care.

As mentioned above, in Kuwait there are substantial limitations on investments made in healthcare research. The main reasons for reduced productivity in health technology are the lack of nationwide developmental plans in the healthcare industry, translational research-focused hospitals, and weak national pharmaceutical and health technological policies, which altogether result in increased dependence on foreign countries for healthcare and medical technology.

Applying the latest knowledge and skills established in medical research is the most effective way to advance healthcare, and to extensively get involved in translational medical research. Building innovative national health research strategies and developing healthcare technology is indispensable to enable Kuwait to be an international hub that delivers advanced healthcare services.

Among the most significant areas that rely on external resources are the medical technology and healthcare sectors. Although a developmental approach is required to tackle nationwide healthcare-related challenges, the health sector of Kuwait is burdened with gaps and limitations. Due to this, the number of patients seeking treatment overseas has increased from 3869 cases in 2013 to 16,085 cases in 2017. In Kuwait, patients with cancer are the predominant population that seeks overseas treatment, followed by 16%, 14%, 11%, and 9% of patients with orthopedic, pediatric, gastrointestinal, and cardiac diseases, respectively.

2. Challenge 1: Inadequate health information and technology

The main reasons for the decline in productivity in the field of health technology are lack of national health industry development plan, lack of a translational research complex hospital, and weak national pharmaceutical and health technology policies, resulting in dependence on foreign countries for health instrument and medical technologies.

Health technology is indispensable for improving health surveillance, increasing health awareness, and helping clinicians make better-informed clinical decisions. For an economy's progress, creating databases containing patient records is crucial, as is maintaining those databases and making them available to promote further research. This would in turn encourage epidemiological studies, ensure proper diagnosis, and improve evidence-based care.

Health technology and medical tourism have been important components of economic growth in many countries, including Korea, India, Thailand, the United Kingdom, the United States, Germany, Singapore, and Japan.

3. Challenge 2: Lack of investment in health research

The investment in health research is meager in Kuwait. Applying the latest knowledge and skills through research is the most effective way to influence health, and it is extensively involved in translational medical research.

Translational research consists of performing novel studies, reporting unprecedented results, and integrating them into clinical practice. Owing to the complexity of current clinical practice and the limited availability of patient information, the application of novel information has become increasingly challenging. Creating strategies for the national health industry and performing a higher number of translational studies will spur health technology innovation in Kuwait.

The purpose of global translational research and policies is to develop creative and innovative healthcare-based knowledge to further enable a nation to become a medical hub and to promote inbound medical tourism around the globe. By building innovative hospitals and developing innovative technology and treatments, Kuwait has the potential to become an international medical hub for outpatient care, along with spending expenditures and recruiting medical tourists from other countries.

Therefore, it has become important to prioritize investment in resources in the health industry and research; the first and most important step is the establishment of a national health research institute. Another key step is to address the measures needed to strengthen health information and data systems.

4. Challenge 3: Lack of infrastructure to support a knowledge-based economy

A knowledge-based economy refers to any nation that is greatly dependent on knowledge- and information-intensive activities to promote technical and scientific advancement. In regard to Kuwait, the development of infrastructure by key stakeholders is vital to ensure advancement of the healthcare industry. Moreover, it would help in supporting local demand by providing the required resources, with a prospect for creating a knowledge-based economy (Kuwait Foundation for the Advancement of Sciences, 2017; Mossialos et al., 2018).

5. Challenge 4: Lack of periodic assessment of population health

The standard practice of periodically assessing the health of the population, employed by most developed countries such as the United States and those in Europe, is conducted every five years with standard templates developed by an expert team. It ensures efficiency, consistency, regularity, reliability, and accuracy, especially in the context of population-based data; thus, evidence-based policies are built on solid, comprehensive, and updated information. In 2009, the first national nutritional survey was conducted in Kuwait to assess the health of the population. Unfortunately, it is the only study conducted on a national scale. In regard to this, a more periodic assessment of population health needs to be conducted in Kuwait to increase epidemiological studies and further improve healthcare practices (Salman et al., 2020a; Powell and Snellman, 2004).

6. Challenge 5: Healthcare Research and Development (R&D) management systems

The role of healthcare R&D management systems is to determine the priority of investment in R&D, such as prioritizing the area and size of investment. Furthermore, this also assists in the establishment of relevant and impactful institutions that conduct R&D. The R&D management system serves to organize and evaluate the performance of R&D, share results, and incorporate relevant results in the overall national development plan. R&D management is also involved in the management of funds and other activities (such as tracking R&D funds, institutes, researchers, educational institutes, and industries) (Salman et al., 2020a; Powell and Snellman, 2004). Thus, the system must be managed by a government entity that can integrate R&D-related information from various fields, and incorporate the data to optimally design national development plans for the country. Currently, only private entities such as the Kuwait Foundation for the Advancement of Science (KFAS) are the major funding agencies for all the conducted research, which puts the burden on employees to keep track of all the applications, data collection reports, and funds from local research institutes. This causes a delay in the publication of data, and ultimately delays the development of national plans and strategies (World Health Organization, 2020; The World Bank Group, 2020).

7. Challenge 6: Lack of information technology records of patients

In Kuwait, around 43% of the hospitals still use paper-based medical records. Not only is this extremely outdated, but it can also cause various other challenges, such as loss of records, writing errors, duplicate data, analysis-related errors, and lack of storage. Electronic health records (EHRs) are tools that enable healthcare specialists to collect, store, and manage the health-related information of patients in an electronic format, where it can be easily accessed and shared between different authorities at any location (Alnashmi et al., 2022; Salman et al., 2020b).

References

Alnashmi, M., Salman, A., AlHumaidi, H., Yunis, M., Al-Enezi, N., 2022. Exploring the health information management system of Kuwait: lessons and opportunities. Appl. Syst. Innov. 5 (1), 25.
Kuwait Foundation for the Advancement of Sciences, 2017. Identifying Priority Sectors in Kuwait. Kuwait Foundation for the Advancement of Sciences, Kuwait City, Kuwait.
Mossialos, E., Cheatley, J., Reka, H., Alsabah, A., Patel, N., 2018. Kuwait Health System Review. London School of Economics and Political Science, London, UK.
Powell, W.W., Snellman, K., 2004. The knowledge economy. Annu. Rev. Sociol. 30, 199–220.
Salman, A., Tolma, E., Chun, S., Sigodo, K.O., Al-Hunayan, A., 2020, September. Health promotion programs to reduce noncommunicable diseases: A call for action in Kuwait. In: Healthcare. vol. 8, no. 3. Multidisciplinary Digital Publishing Institute, p. 251.
Salman, A., Fakhraldeen, S.A., Chun, S., Jamil, K., Gasana, J., Al-Hunayan, A., 2020, September. Enhancing Research and Development in the Health sciences as a strategy to establish a knowledge-based economy in the State of Kuwait: a call for action. In: Healthcare. vol. 8, no. 3. Multidisciplinary Digital Publishing Institute, p. 264.
The World Bank Group. 2020 Researchers in R&D (per Million People). Available online: https://data.worldbank.org/indicator/SP.POP.SCIE.RD.P6 (Accessed on 17 April 2020).
World Health Organization, 2020. Kuwait Noncommunicable Diseases Profile. Available online: https://www.who.int/nmh/countries/kwt_en.pdf?ua=1 (Accessed on 12 May 2020).

Index

Note: Page numbers followed by f indicate figures, t indicate tables and b indicate boxes.

Musculoskeletal radiographs (MURA) dataset, 97
MyAudio2Go.com, 8–9

N

National Board of Medical Examiners (NBME), 198
National Board of Osteopathic Medical Examiners (NBOME), 198

O

Optical character recognition (OCR), 8–9

P

Partial least square (PLS) method, 143–144
Performance expectancy (PE), 138–139, 150–151*t*
Personal innovativeness (PI), 141–142, 150–151*t*
Physiotherapy, 79, 89–90
Privacy and security
  access control systems, 66–67
  trust management, 65–66
Prosthetics, 80–81

R

Remote patient monitoring (RPM), 52–53, 52*f*
Research and development (R&D), 196
Research project, for DT in SCFHS
  objectives, 186
  research questions, 186
    business process modeling and reengineering, 186–187
    digital innovation, 189
    platform-mediated integrated approach, 188
    policy-making recommendation, 188–189
    quality frameworks and KPIs, 187–188
    technology-driven value proposition, 187
Reverse transcription polymerase chain reaction test (RT-PCR), 95
ReWalk framework, 88
Robotic rehabilitation, 79, 82–83
Robotics, 5–7
Roomba, 6–7
RPM. *See* Remote patient monitoring (RPM)
RUPERT, 85

S

Saudi Commission for Health Specialties (SCFHS)
  digital transformation (DT)
    accreditations, 198
    admission, 197
    blockchain, 179
    classification and registration, 198
    continuing professional development, 199
    councils, 199
    exams and evaluation, 197

functional view, 195
health workforce planning, 181–182*t*, 199
indicative use cases, 195–203, 199–200*t*
literature review on, 175–181, 175–178*t*
managed service providers (MSPs), 180
matching system, for selecting postgraduate trainees, 196
mobile health apps, 179
platform-mediated ecosystem, 180
research challenges, risks, and milestones in, 189–190, 189–190*t*
research project for, 186
SCFHS matching system (SCFHS-MS), 196
strategic framework for, 182–185, 191
training design, development, and implementation, 196–197
value chain, 180
value proposition, 179–180
profile of, 173–174
Saudi Data and Artificial Intelligence Authority (SDAIA), 28, 168–169
Saudi healthcare system
  action plan and milestones, 214–215
  digital transformation, 208, 210*f*
  e-health transition, 212*f*
  interrelationship of solution classes, 213*f*
  roadmap for strategic transition, 211–215
  telehealth applications, 209*f*
Saudi Health Informatics Competency Framework (SHICF), 206–207, 207*f*
Saudi health sector transformation program, 222
Scalable intangibles, 125–127
SCFHS. *See* Saudi Commission for Health Specialties (SCFHS)
SCFHS matching system (SCFHS-MS), 196
Scopus database, 225–226
Self-determination theory, 137
Services master online system (COSMOS), 8–9
Skype, 14
Smart contracts, 61–62
Smart devices, 133
Smart health technologies, 53
Smartphone, 5–6, 135
Social influence (SI), 139–140, 150–151*t*
SPRP20 strategic plan, 217–218
Standard healthcare network, 123*f*
Standard wars, mobile technologies, 5–6
Strategic framework, for DT in SCFHS, 184*f*
  functional aspects, 182, 183*f*, 191
  strategic components and impact dimensions, 182–183, 183–184*f*, 191
    digital innovation, 183–184
    platform-mediated ecosystems, 185

Printed in the United States
by Baker & Taylor Publisher Services

Printed in the United States
by Baker & Taylor Publisher Services